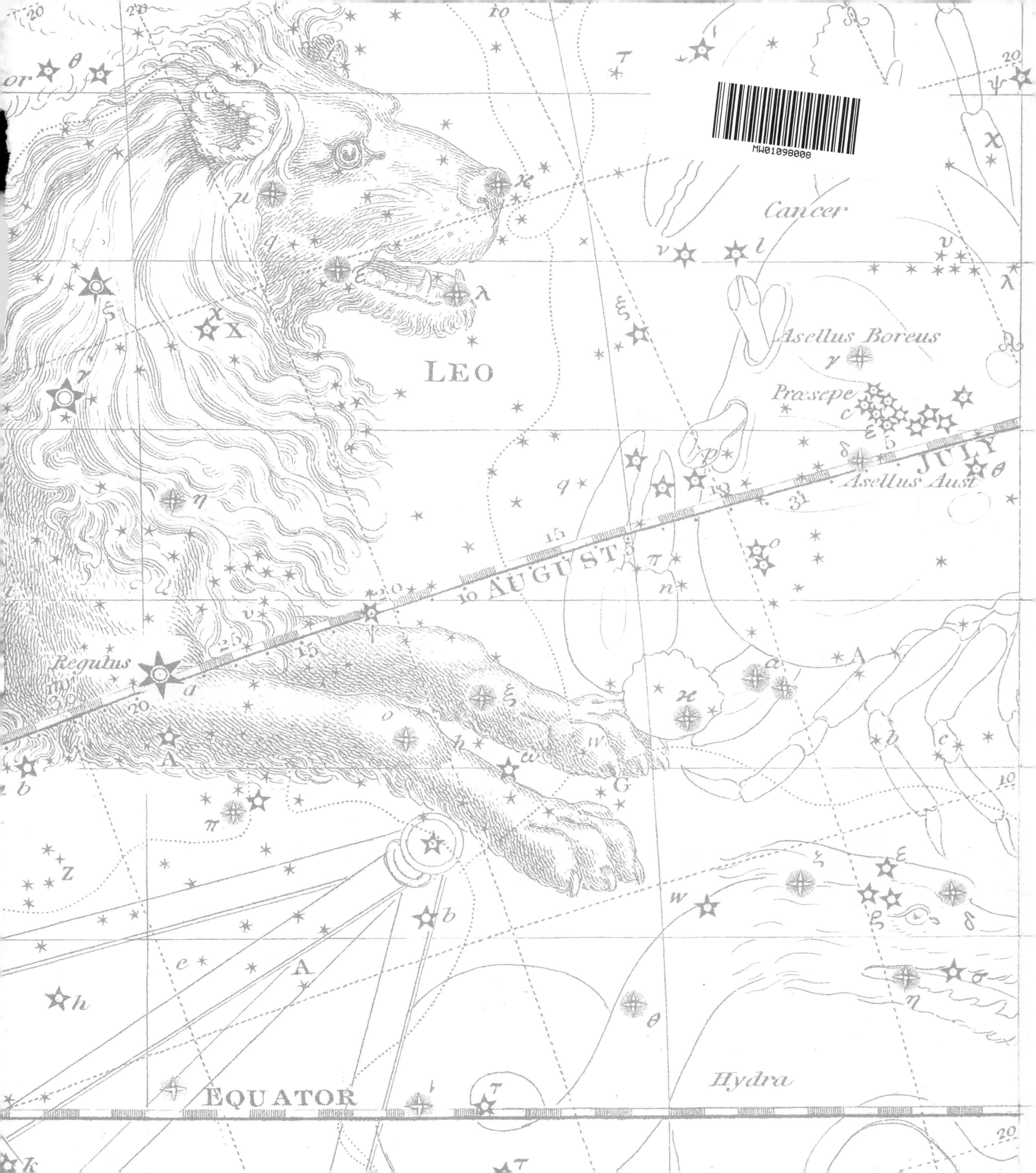
LEO
Cancer
Asellus Boreus
Præsepe
Asellus Aust
JULY
AUGUST
Regulus
EQUATOR
Hydra
MW01098008

CELESTIAL ATLAS

A Journey in the Sky through Maps

Elena Percivaldi

CELESTIAL ATLAS

A Journey in the Sky through Maps

Contents

Clockwork Model of the Solar System

This illustration taken from Smith's Illustrated Astronomy *by Asa Smith (Boston, first edition, 1848) shows students how the 'orrery' or clockwork model works. The book was published as a teaching aid in public schools.*

LEVERRIER
JUNO
VESTA
PALLAS
CERES
MARS
EARTH
JUPITER
VENUS
ASTRÆA
MERCURY
S
COMET
SATURN
HERSCHEL
JUNE
JULY
AUG
SEP
OCT
NOV

Introduction

Understanding and depicting the secrets of the cosmos is a goal that Man has set for himself since the remote past. In the Near East the first who sought to achieve this were the populations that lived in the Fertile Crescent, beginning with the Sumerians, who in the IV Millennium BC identified the star alignments, giving them a name and a function. This led to the birth of the twelve constellations of the Zodiac, which are quite similar to those of the present day. The notions that the Sumerians elaborated were passed on to the Assyrians and Babylonians, and then to the ancient Egyptians, eventually arriving in Greece through Eudoxus of Cnidus (5th-4th century BC), a pupil of Plato, who studied at Heliopolis. His writings were set in verse by Aratus of Soli (4th-3rd century BC) in *Phaenomena*. Thanks to this successful work, which was translated into Latin and commented on by various Latin authors – including Varro Atacinus, Cicero, Germanicus and Avienus – the concepts were bequeathed to the Middle Ages and continued to circulate – albeit at time revised in accordance with Christian theology – up to the beginning of the Renaissance.

The astronomer who attributed a form and myth to each constellation was Eratosthenes of Cyrene (3rd-2nd century BC) in his *Catasterisms*. This work was taken up in the 2nd century in *De Astronomiae*, attributed to the Latin author Hyginus, but in the meantime the body of stories elaborated around the stars was enhanced by influences stemming from epics, the myths and legends in the *Bibliotheca* of the pseudo-Apollodoro, classic works such as the *Argonautica* by Apollonius Rhodius, and lastly by the verses of Ovid, who in his *Metamorphoses* had imparted poetic vigor to the gods' transformations of heroes and heroines into plants or animals. This immense treasure transmitted dynamism and inspiration to the European and Arab populations for almost 1500 years, from the Middle Ages to the modern age. However, there was something that the Middle Ages did not know. In the 2nd century Claudius Ptolemy had proposed the model of a geocentric universe and had described 48 constellations in his *Almagest* (the original title was *Mathematical Treatise*, later known as the *Great Treatise*). This work, unknown in Europe due to insufficient knowledge of ancient Greek, was translated into Arabic in the 9th century with the title *Al-magist* and integrated with the names of the stars. When in the late 12th century the treatise was finally translated into Latin (the famous *Almagestum*), it sparked interesting repercussions; although the constellations referred to ancient Greek mythology, they were given Latin names, while the stars became popular with their Arabic names.

Illustrating the sky was by no means an easy task. The first persons to attempt this feat were, once again, the ancient Greeks, who in the 2nd century BC created a

Pisces

The zodiacal constellation Pisces in the oldest manuscript (9th century) containing Cicero's Aratea, *the Latin version of* Phaenomena *by Aratus of Soli: in the drawing is an excerpt from* De Astronomiae *by the pseudo-Hyginus, which describes the constellation (British Library, Ms Harley 647, f. 3v).*

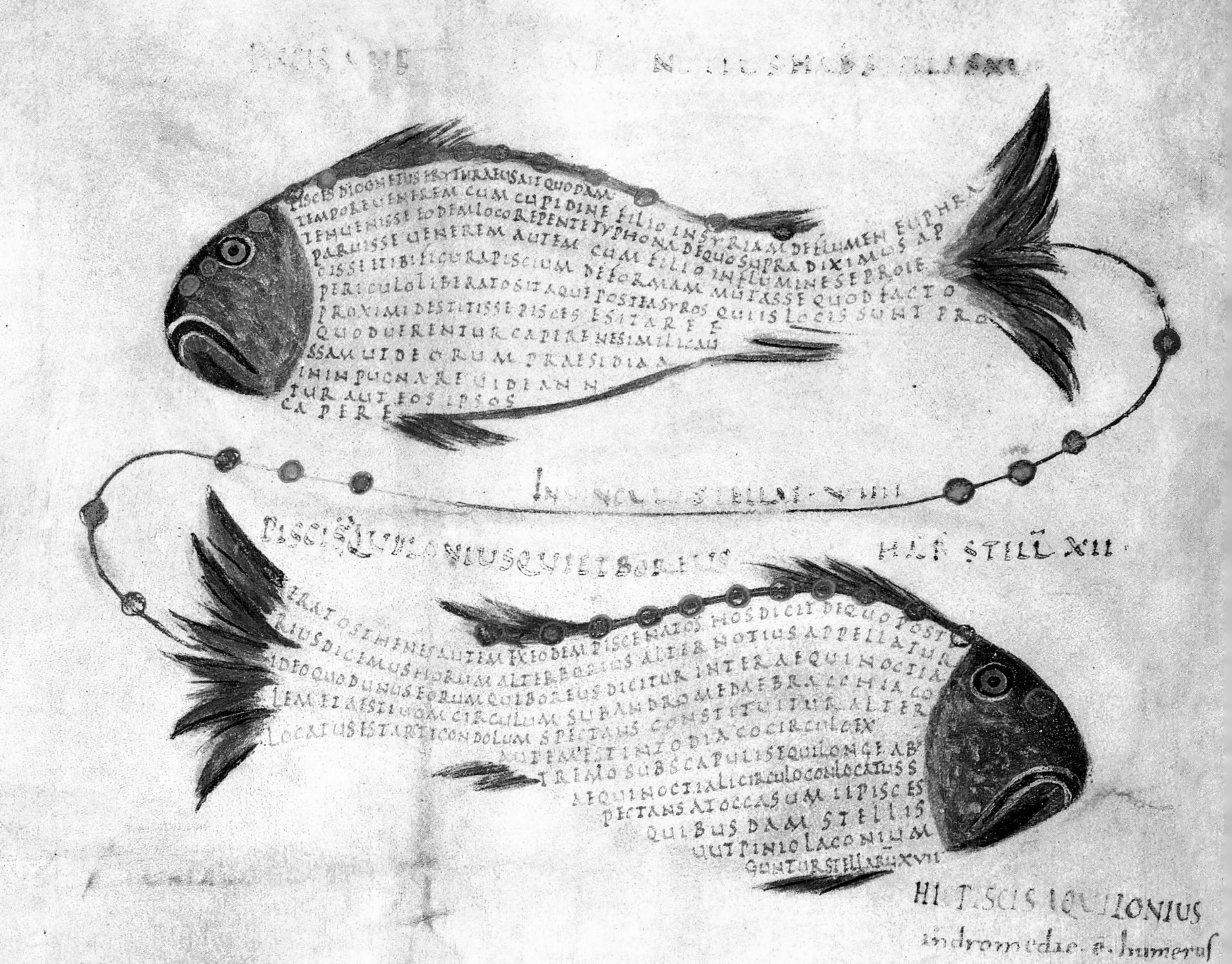

Pisces quorum alter paulo praelabitur ante
Et magis horrisonis aquilonis tangitur alis
Atque horum e caudis duplices uelut esse catenae
Dices quae diuersae per lumina serpunt
Atque una tamen in stella communiter haerent
Quem ueteres soliti caelestem dicere nodum
Andromedae leuo ex umero si quaerere perges
Adpositum poteris supra cognoscere piscem

PISCES

sculpture of Atlas (the Roman copy of which is known as the *Farnese Atlas*) bearing the celestial sphere. The first planispheres, on the other hand, were realized only in the 9th century by the Arabs, while as regards the constellations the important figure – besides the above-mentioned Aratus, who continued to be mentioned in innumerable illuminated manuscripts – is Al-Sufi, who wrote *The Book of Fixed Stars* (AD 965), in which refined drawings accompanied the descriptions. Throughout the Middle Ages and up to the 17th century star maps continued to be produced, with more attention paid to the aesthetic aspect than to scientific accuracy. For example, the position of the stars was at times adapted to the anatomical details of the constellations. Something began to change in 1540, when the German humanist and cosmographer Peter Apian published *Astronomicum Caesareum*, a treatise that illustrated Ptolemy's theories with the aid of rotating paper volvelles, thus combining the pursuit of beauty with renewed scientific rigor.

Taula de la Disposicio de las Esperas del IIII Elementis e dels VII Planetas e del Cel Estelat

This highly colored 14th century miniature from Matfre Ermengau de Bézier's Le breviari d'amor *(British Library, Ms. Royal 19.C.I., f. 50r) represents the cosmos according to Ptolemy but in a Christian key. In fact, four angels are turning the spheres.*

Catalan Atlas

The spectacular second sheet of the Catalan Atlas *(Bibliothèque nationale de France, ms. Espagnol 30), dating to c. 1375, depicts the solar and lunar calendar accompanied by the planets known at the time (as well as the Sun), which are arranged in concentric circles around the Earth.*

And it is precisely at this point, with reproductions of Apian's ingenious volvelles, that we set off into the golden age of cartography. It begins in 1603, with the publication of *Uranometria* by Johann Bayer, who inserts the twelve constellations of the Southern Hemisphere, which had been introduced only a few years earlier (1598) by the Dutchman Petrus Plancius, thus filling in a portion of the sky that until then had remained obscure since it was unexplored. From an iconographic standpoint, the constellations in Bayer's sky were figures taken from Greek mythology. In 1627 his colleague Julius Schiller, on the other hand, embarked on the bold operation of 'rechristening' them by substituting the pagan heroes with the protagonists of the Old and New Testament. His experiment was not very successful, but it has been included in this volume for its marked originality.

A few decades later the first great masterpiece of celestial cartography saw the light of day: Andreas Cellarius's *Harmonia Macrocosmica.* Next came the work of Johannes Hevelius, *Prodromus Astronomiae*, which has 56 superb maps that introduce eleven more new constellations. Then it was the turn of the Englishman John Flamsteed, Astronomer Royal at Greenwich and author of the *Atlas Coelestis* (1729), with the most complete, as well as the absolutely most impressive, star catalog produced up to that time. It was followed by Johann Gabriel Doppelmayr's *Atlas...* (1742), whose exceptional plates, which in many editions were colored by hand, certainly justify the ample space they occupy in this book.

This golden age ended with Johann Elert Bode's *Uranographia* (1801), which was the scientific apogee of star maps, containing more than 100 constellations and 17,000 stars, as well as the 'swan song' of artistic cartography. Telescopes were becoming more and more powerful, thus making it impossible to depict all the stars on traditional maps. Therefore, the last mythological atlases – the Scot Alexander Jamieson's *Celestial Atlas* (1822) and the *Geography of the Heavens* (1833) by the American Elijah Hinsdale Burritt, as well as the popular series of star chart cards in *Urania's Mirror* (1824) – focused on astronomy enthusiasts, simplifying the subject matter and contributing to its dissemination, while abandoning all pretense of providing a precise and up-dated scientific treatise.

The 1922 meeting of the International Astronomical Union (IAU) established the number of accepted constellations at 88, which are still official. Eight years later, Eugène Delporte's *Délimitation Scientifique des Costellations* determined their boundaries and marked the definitive end of the epoch of myths and inaugurated modern cartography.

Zubdat al-Tawarikh

A celestial map with lunar mansions and the signs of the Zodiac, from a manuscript of the Zubdat al-Tawarikh *(Cream of Histories) by Seyyid Loqman Ashuri (Istanbul, Museum of Turkish and Islamic Art). This codex was created in 1583 for Sultan Murad III and is certainly one of the greatest achievements of Ottoman miniature production.*

الميزان
العقرب
القوس
الجدي
الدلو

Peter Bienewitz Apian

(1495-1552)

Peter Bienewitz was born in Leisnig, Saxony to a family of humble origin but well to do. His father was a shoemaker who enrolled Peter in the Rochlitz School, where he could learn the rudiments of Latin, and then in the University of Leipzig. At this latter the young man came into contact with the lively intellectual circles of the city and adopted the custom – which was quite popular among the humanists of the time – of Latinizing his family name, changing the original Bienewitz ('of the bees') into the classicizing Petrus Apianus.

In 1519 Apian enrolled in the University of Vienna to study with the great mathematician and astronomer Georg Tannstetter in order to increase and perfect his scientific and geographic learning. However, only two years later he had to leave the city to escape from an epidemic of the plague, which devastated Vienna in 1521, taking refuge at Regensburg and later in Landshut, in Bavaria. In this latter city Peter was able to publish his first work, Cosmographicus liber *(1524), in which he gave a precise description of the constellations and the Earth's surface that were known at the time. This work, which contained numerous large plates, was a huge success, reprinted at least thirty times and translated into fourteen languages, becoming a 'classic' of its kind up to the end of the 16th century. The economic success of his treatise allowed Peter to marry Katharina Mosner, the daughter of one of the members of the City Council. The couple had fourteen children, one of which, Phillip Apian, would follow in his father's footsteps and become a leading cartographer.*

In the following years, Peter Apian's prestige and fame extended well beyond the confines of Landshut, embracing all of Bavaria. In 1527, he was offered a post at the university in nearby Ingolstadt; he accepted, but did not want to limit his work to teaching mathematics. Germany in that period was in a state of agitation because of the passionate sermons of Martin Luther, the father of the Protestant Reformation. In Bavaria, which had remained Catholic, Apian opened a print shop that concentrated solely on publishing quality geographic and cartographic works, as well the treatises of Johann Eck, one of Luther's most important and most polemical antagonists.

The definitive turning point for Apian took place in 1530, when Emperor Charles V publicly lauded the Cosmographicus liber *at the Imperial Diet of Augsburg. Having obtained the good graces of the emperor, Apian would soon be entitled to display heraldic arms. Apian repaid this attention by dedicating his most famous and important work,* Astronomicum Caesareum *(1540), to Charles V. The emperor in turn reciprocated by gifting him with a large sum of money, then making him his Imperial Mathematician and Knight, and lastly granted him the title of Count Palatine. On the strength of all this 'imperial esteem', Apian was sought after by the most prestigious universities in Europe, such as Vienna, Leipzig, Padua and Teubingen, but he preferred to remain in 'his' Ingolstadt until he died at the age of 57, on 21 April 1552.*

Portrait of Peter Apian in an engraving from Justin Windsor's Narrative and Critical History of America *(1884).*

PETRUS APIANUS BINUITIZIUS ASTROLOGUS
QUOD COELOS PANDIS, RADIO QUOD SIDERA PINGIS
GERMANUS MERITO DICERIS ARCHIMEDES,

Astronomicum Caesareum

(1540)

The *Astronomicum Caesareum* was first published at Ingolstadt in 1540. This treatise is a survey of the geographic and astronomical knowledge of the time that was not very original, since it was based on Ptolemy's theories, presented without any significant alterations. Indeed, what makes this work so outstanding are the numerous engravings that accompany the text, and above all the 34 beautiful paper wheel charts known as volvelles, with rotating and superposed parts, which were invented to compute the phases of the moon and the terrestrial and astronomical distances, as well as predict the position of the planets. The volume was dedicated to Charles V, Emperor of the Holy Roman Empire, and to his brother Ferdinand I of Habsburg. As examples to make a calculation of lunar eclipses, Apian chose phenomena that occurred in conjunction with certain significant dates in the life of the two monarchs; the partial eclipses of 15 November 1500 (the year Charles V was born) and 15 October 1503 (a short time before the birth of Ferdinand), and the total eclipse of 6 October 1530, the year the emperor was crowned. The latter's appreciation of this work was certainly partly responsible for the success of this work, which went through many editions and reprints and was considered a classic until end of the 16th century, when it was superseded by more up-dated and complete works, first and foremost the atlas of Andreas Cellarius.

Frontispiece

The frontispiece of Apian's Astronomicum Caesareum *represents the Sun, on which a large dragon seems to be lying, the symbol of the ascending and descending nodes. The title is written in Gothic lettering; this is the famous Textura style, used for university and philosophical works and chosen by Gutenberg for the first printed edition of the Bible, which enjoyed immediate popularity throughout Germany.*

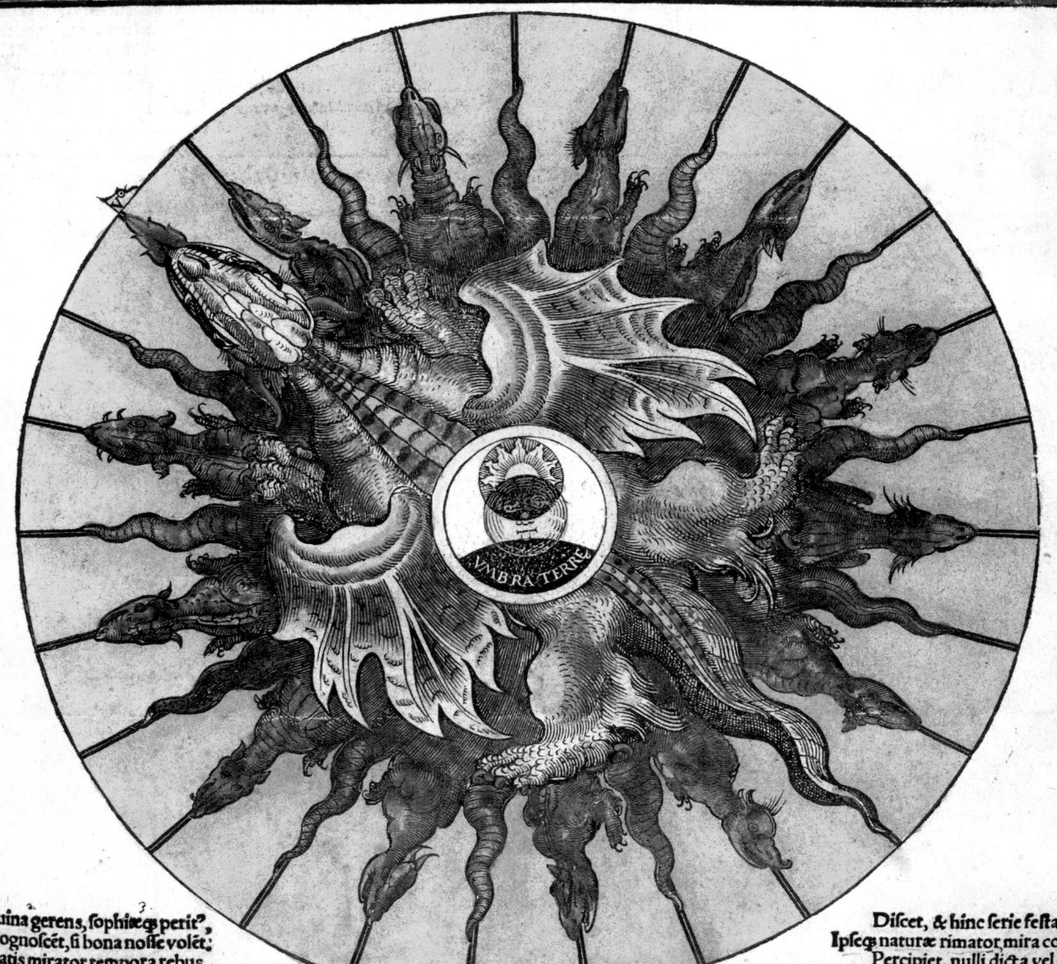

Historicus, diuina gerens, sophiæq; perit',
Hic sua cognoscet, si bona nosse volet;
Namq; vetustatis mirator, tempora rebus
Distribuet; verè dum canet historias.
Ipse sacri præses, noctes æquare diebus
Discet, & hinc serie festa locare sua;
Ipseq; naturæ rimator, mira cometæ
Percipiet, nulli dicta vel acta prius;
Sed caueant animis adsint liuore perustis,
Hæc etenim labes cernere vera neget.

Ad Principem Thomam Thyrlibeum Episcopu Vistmonasteriense

Imagines Syderum Coelestium

In this beautiful paper wheel chart (volvelle), Apian illustrates all 48 constellations in keeping with Ptolemy's classification. The volvelle, a mobile paper disk, is placed over a plate that has a series of graduated nicks and scales along the edge of the circumference: when it is turned it can make a precise calculation of the precession of the equinoxes. This image, in which the utmost attention is paid to every detail, was enormously successful and was even copied and hand-colored by an anonymous author a few years later. The chart is now kept in the Rein Abbey, in Austria.

The Eclipse of the Moon

This plate, which is also a volvelle, allows one to calculate and predict eclipses. The central disk representing the Moon with a human face can be turned by means of a red thread connected to the middle of the volvelle. Note the supernatural hand that, as in the other plates, appears from the Heavens to hold up the entire diagram.

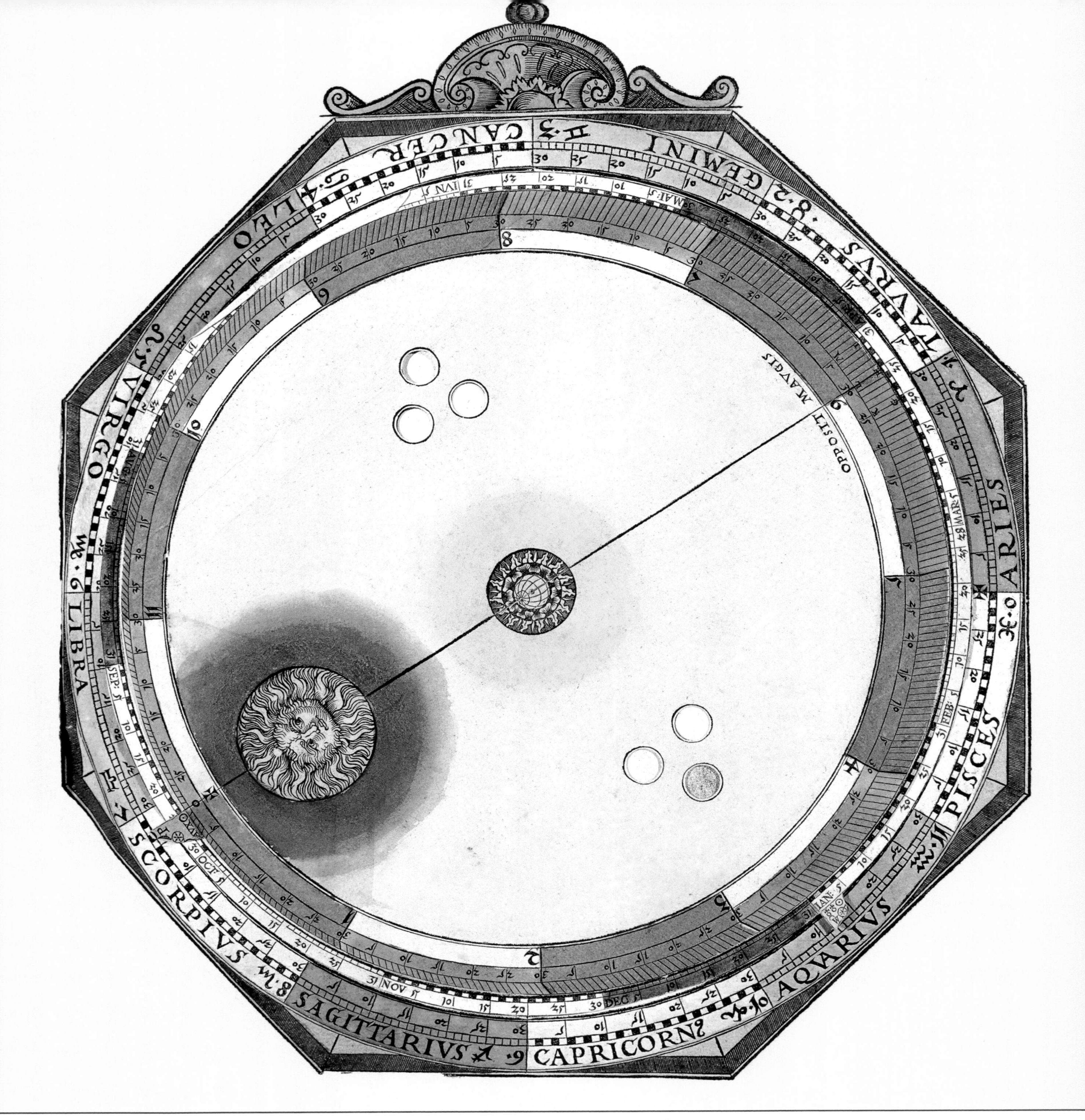

Representation of the Movement of the Sun

Here the protagonist of the volvelle is the Sun as it orbits around the Earth. Apian was a proponent of the geocentric system: De revolutionibus orbium coelestium, *in which Nicolaus Copernicus anouunced his revolutionary heliocentric theories, would be published only in 1543, that is, three years after the publication of* Astronomicum Caesareum. *Apian probably already knew of them but ignored them, perhaps in order to protect his position as Charles V's court mathematician.*

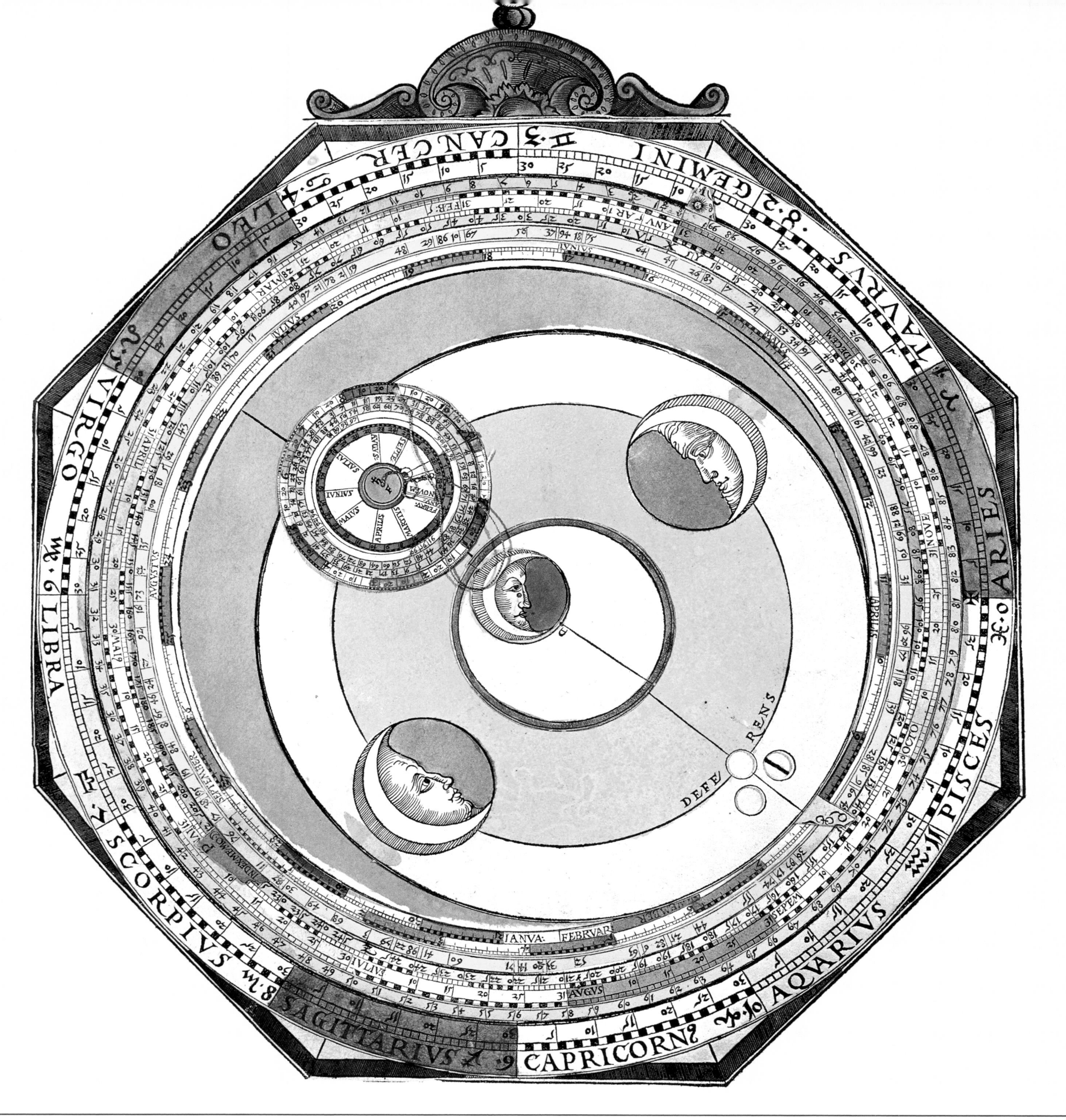

The Movement of the Moon

This plate features the movement of the Moon and comprises various volvelles, one on top of the other. The main one shows the movement in the celestial quadrants and their respective phases, while the smallest one – which in turn consists of superposed wheels of different colors that can be turned with the green threads (a minor masterpiece of miniature graphics and prints) – was used for calculating the position of the Earth's satellite every day of the year.

Johann Bayer
(1572-1625)

We know very little about Johann Bayer – also called Giovanni Baiero, following the humanists' custom of 'Latinizing' their names – except that he was born in Rain, in southern Bavaria, in 1572 and that he was an illustrious lawyer in Augsburg before becoming an astronomer. He had many other passions, such as archaeology and mathematics, but is famous as the person who produced the first star atlas ever to deal with the celestial sphere in its entirety. The basis of this work, titled Uranometria *and published in 1603, are the studies carried out by a contemporary, the Danish astronomer Tycho Brahe, whose star chart (which included 1005 stars all told) circulated at that time only in manuscript form or was partly reproduced in the beautiful celestial globes of the cartographers Petrus Plancius, Hondius and Willem Blaeu. In fact, Brahe's work was published in tabular form only in 1627, in Johannes Kepler's* Tabulae Rudolphinae. Uranometria, *which comprises more than 1200 stars (the exact total number is disputed since many stars lack a name), is an even more complete catalog because, besides the 48 already known 'classical' Ptolemaic constellations, it also presented, for the first time, the twelve Southern Hemisphere constellations that were unknown to most Europeans: the representations of the twelve were based on the observations of the Dutch navigator Pieter Dirkszoon Keyser (also known as Petrus Theodorus), who in turn had corrected the estimates formulated previously by the Florentine explorers Amerigo Vespucci and Andrea Corsali.*

Bayer himself measured the size of the stars for his Uranometria, *and he introduced Greek letters to label the brightest stars. Before Bayer, a scholar from Siena, Alessandro Piccolomini, had conceived a similar procedure of designating the stars in his treatise* De le stelle fisse *(1543), but he used Latin instead of Greek letters. However, Bayer's system was adopted and in fact is still used. In Bayer's Nomenclatura the Latin name of the constellation the star belongs to is placed after the Greek letter and is declined in the genitive case. For example,* Alpha Centauri, *or* α Centauri, *indicates the brightest star in the Centaur constellation.*

A curiosity: as is often the case, the constellations on celestial maps are accompanied by a drawing depicting the person, animal or object after which they are traditionally named. However, in Bayer's Atlas human figures such as Orion are represented as seen from behind, while they are commonly portrayed frontally, facing the Earth. The reason why Bayer made this rather strange decision is still unknown and most probably will remain a mystery.

Johann Bayer died on 7 March 1625 in 'his' Augsburg. A crater on the Moon is named after him.

Boötes

This plate depicts the constellation of Boötes, the plowman or herdsman, seen here with a sickle in his hand rather than a club, spear or staff, which are equally widespread in the iconography of this mythical figure. The principal star is Arcturus (Alpha Boötis), which can be seen on the left knee.

k
i
θ
g
κ
h
φ
ν
β
λ
M
μ
δ
γ
N
A
χ
b
ψ
ρ
σ
ε
c
ω
d
ξ
f
α
ο
e
π
ζ
η
τ
υ
P

Uranometria

(1603)

The complete title of this work is *Uranometria: omnium asterismorum continens schemata, nova methodo delineata, aereis laminis expressa* (or "*Uranometria*, containing the charts of all the constellations, drawn with a new method and engraved on copper plates"), and it was published at Augsburg in 1603 by Christoph Mang, who made it the pride and joy of his print shop. The term *uranometria* is a neologism that derives from Urania, the Muse of the heavens (*uranos* = sky) and *metria* or measurement: just as geometry is the science that measures the Earth, *uranometria* is the discipline that calculates the extension of the sky.

Bayer's treatise, the first in which the cartographic section was longer than the text, consists of 51 star maps engraved on copper plates by the German Alexander Mair, one for each of the 48 known Ptolemaic constellations, plus a supplement with the twelve southern constellations unknown to Ptolemy. This novelty was nothing less than sensational, because up to that moment celestial maps had had a gap in the central region of the Southern Hemisphere, and now, thanks to the observations of explorers such as Pieter Dirkszoon Keyser and Cornelis de Houtman, the map could finally be considered complete.

Bayer's extraordinary work ends with the two recapitulatory planispheres of the sky of both hemispheres, the *Synopsis coeli superioris borea* (northern sky) and *Synopsis coeli inferioris austrina* (southern sky). These two plates drew inspiration from the engravings executed in 1515 by the great Albrecht Dürer and, like all the other plates, were created with extreme technical prowess and remarkable aesthetic taste and elegance.

Frontispiece

The frontispiece of Uranometria *has numerous allegories that are explained by the inscriptions along the ledges and on the pedestals of the elaborate Baroque architecture. Above, in Greek are* Oudeis eisito *and* Ageometretos, *an adaptation of the motto at the entrance to Plato's school: "Let no one ignorant of geometry enter here." In the middle is a dedication to Eternity, and below are statues of Atlas and Hercules, respectively the teacher and disciple of Astronomy.*

ΟΥΔΕΙΣ ΕΙΣΙΤΩ
ΑΓΕΩΜΕΤΡΗΤΟΣ
AETERNITATI.
IOANNIS BAYERI
RHAINANI I.C.
VRANO-
METRIA,
OMNIVM ASTERISMORVM
CONTINENS SCHEMATA,
NOVA METHODO
DELINEATA,
AEREIS LAMINIS EXPRESSA.
ATLANTI
VETVSTISS.
ASTRONOM.
MAGISTRO.
HERCVLI
VETVSTISS.
ASTRONOM.
DISCIPVLO.
MDCIII

Pegasus

In this plate Bayer represents Pegasus, the winged horse born from the blood of the Medusa. This figure is connected to very ancient myths and probably originated in Asia Minor. It is based on the idea of transfiguring the strength and vitality of a splendid animal like the horse into pure nimbleness and grace, lending it the prerogative of moving freely in the heavens. The name Pegasus derives from the Greek pege, 'spring' or 'fountain'. According to legend, one day this animal flew over Mount Helicon and hit a rock with its hoof, thus creating the Hippocrene fountain, which from then on quenched the thirst of the Muses. This is the reason why Pegasus is also associated with the idea of artistic inspiration, which suddenly springs up, allowing artists to create freely.

Centaurus

Centaurs are the most familiar creatures in classical mythology. Their body is half human and half horse, and they are irascible and wild, very skillful archers and able to attack their enemies with a club. The constellation Centaurus, one of the largest and brightest in the southern sky, is dedicated to one of them, Chiron. In Bayer's plate Chiron is seen while killing the Lupus (Wolf), the constellation next to Centaurus, with a spear. This feature immediately distinguishes him from Sagittarius, the centaur constellation that is part of the Zodiac, since the latter is always depicted armed with a bow and arrow. The brightest star, Alpha Centauri, is located on the left forehoof.

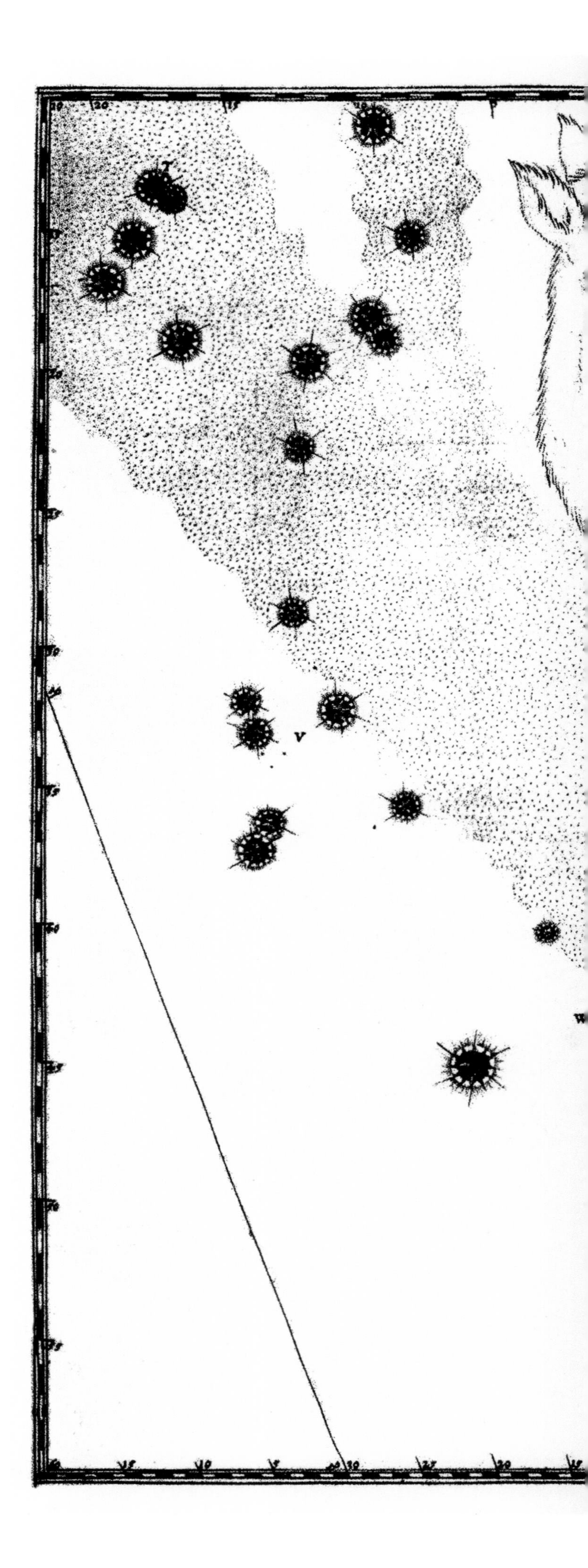

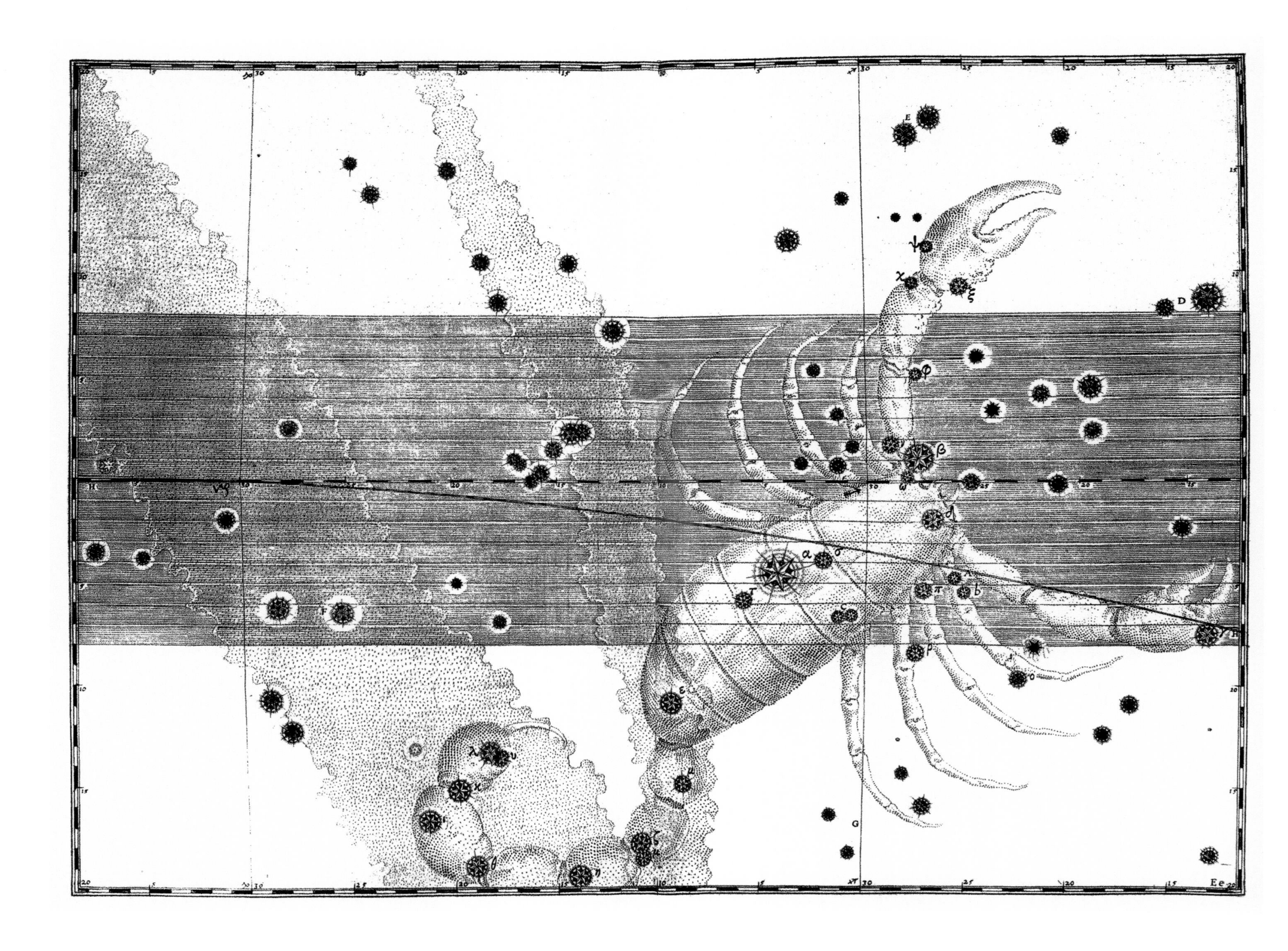

Scorpius

Like all the constellations, Scorpius or Scorpion also has stars of different sizes, according to their relative magnitude. The position of the stars is established in keeping with the calculations of Tycho Brahe. To the left of the last part of the scorpion's tail there is another, almost imperceptible star that Bayer drew but then immediately erased. This is probably the M7 open cluster, also known as the Ptolemy Cluster, which already in the 2nd century AD the Greek astronomer had classified as a "nebula following the sting of Scorpius." Although it is not a star, Bayer decided to represent it all the same, but made it more evanescent so that it would not be confused with the other true stars.

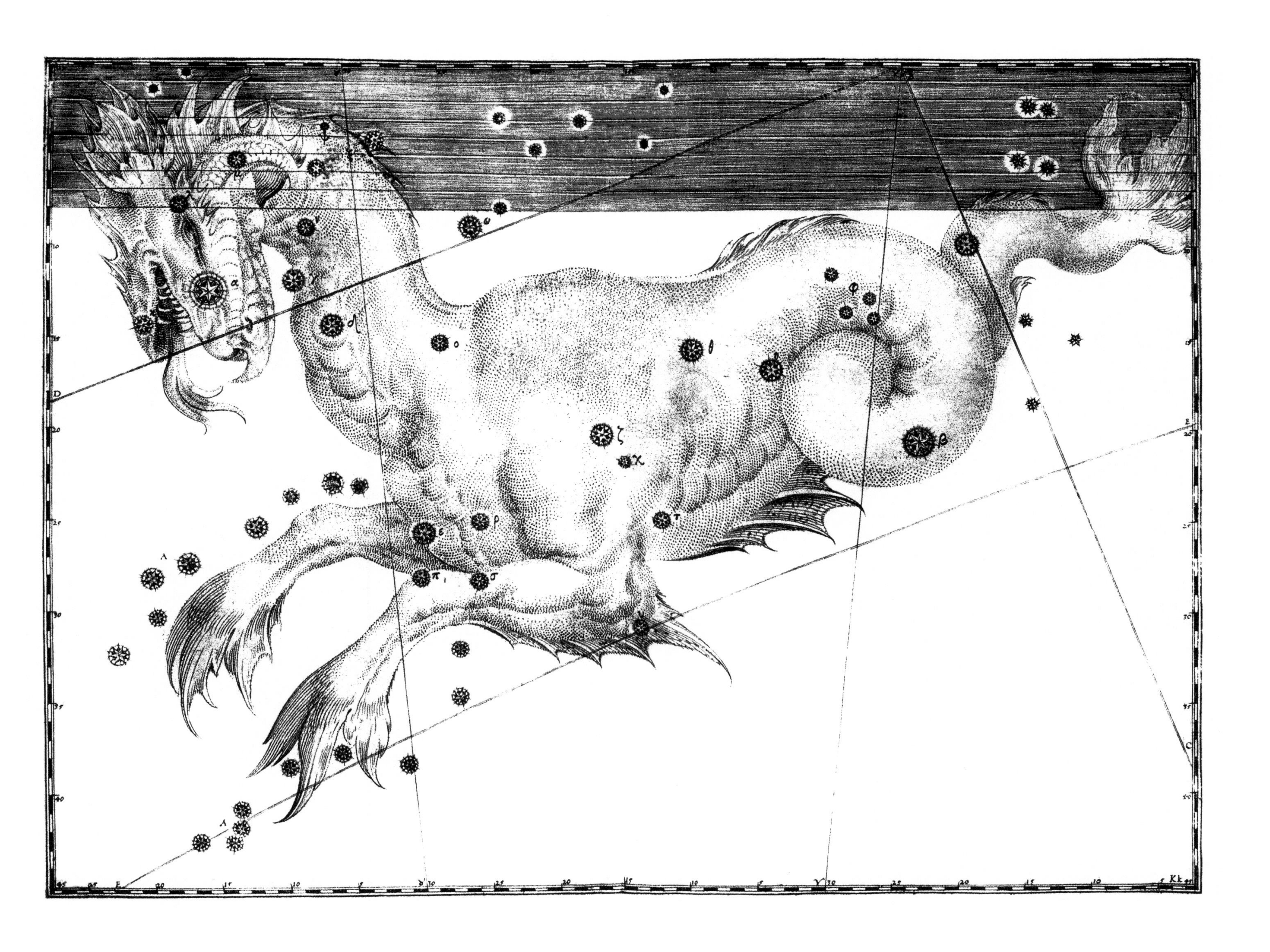

Cetus

In Greek mythology, Cetus was the sea monster that Poseidon had send to plague the seas of the Ethiopian kingdom ruled by Cepheus, as punishment for the arrogance of the monarch's wife, Cassiopeia, for having said she was more beautiful than the Nereids. In order to placate the god, an oracle told the king that he would have to sacrifice his young daughter Andromeda, so Cepheus tied her to a cliff over the sea and left her at the mercy of the hungry monster. However, the hero Perseus killed the terrible creature with his sword, freed the girl and then married her. Cetus, or Whale, is a constellation in the southern sky that lies next to other constellations associated with water, for example Aquarius, Pisces and Eridanus. In the plate, Bayer depicts the monster as a sort of aquatic dragon with a long fiery tongue.

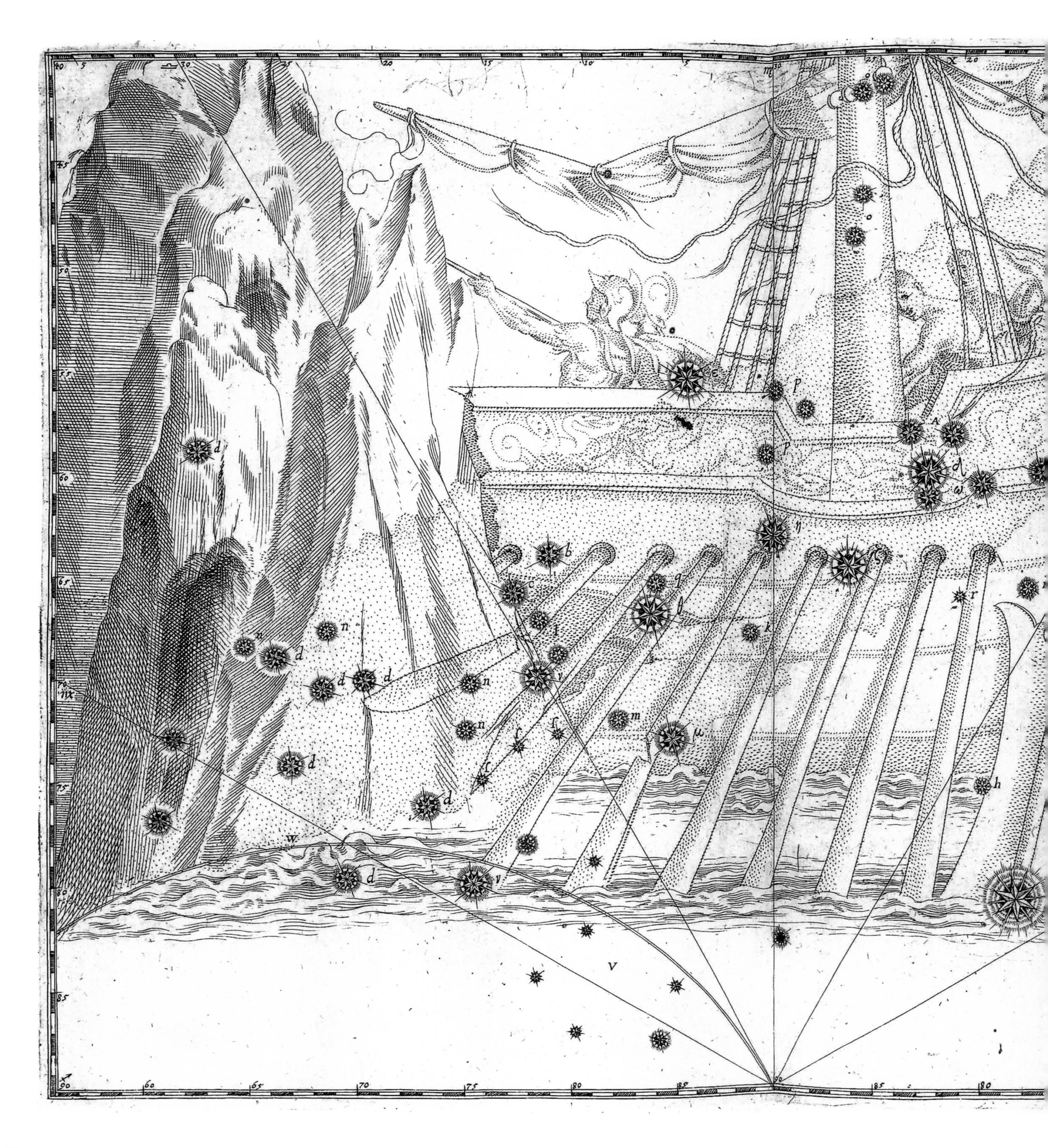

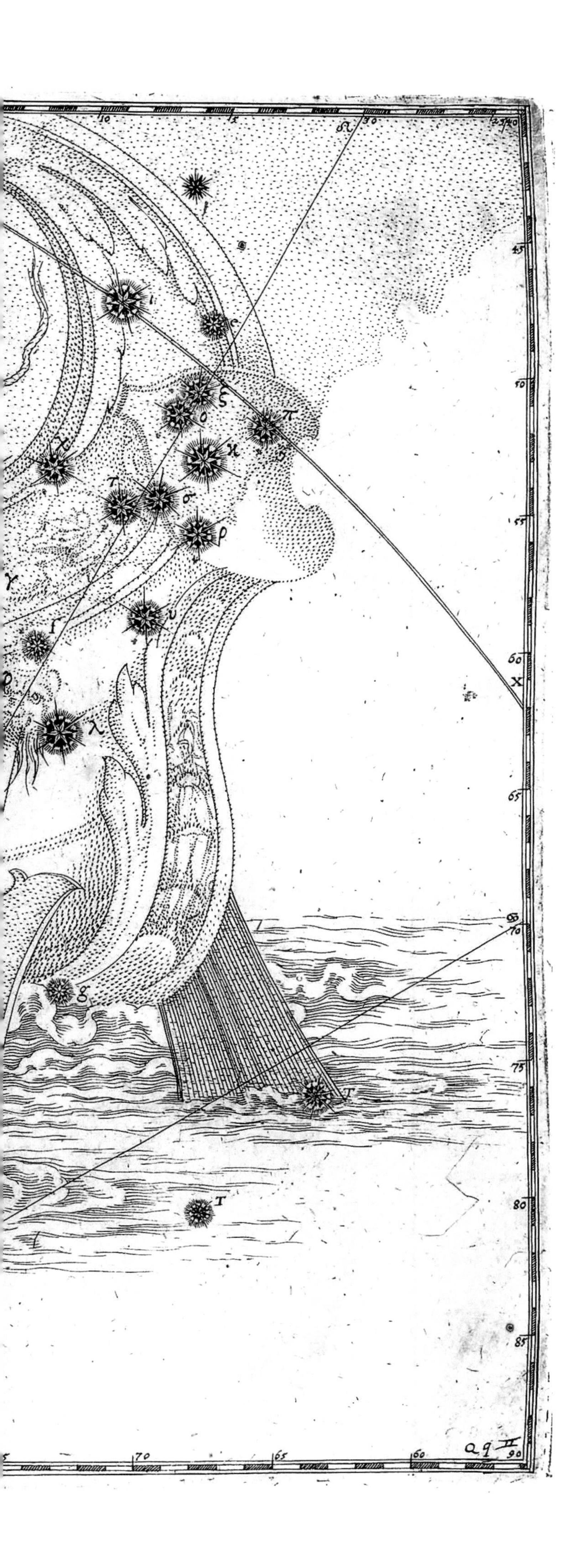

Argo Navis

The Ship Argo (Argo Navis in Latin) was a southern sky constellation that depicted the ship of the same name used by Jason and the Argonauts to sail toward Colchis to find the Golden Fleece. This was considered the largest constellation in the heavens until in the 18th century the French astronomer Nicolas-Louis de Lacaille divided it definitively into four different constellations: Pyxis (Compass), Carina (Keel), Puppis (Stern) and Vela (Sails). In this plate the entire complex – with the exception of Pyxis – is still intact, since we see the entire ship. The stars are indicated with the nomenclature introduced by the astronomer, which is based on Greek letters in decreasing order of magnitude.

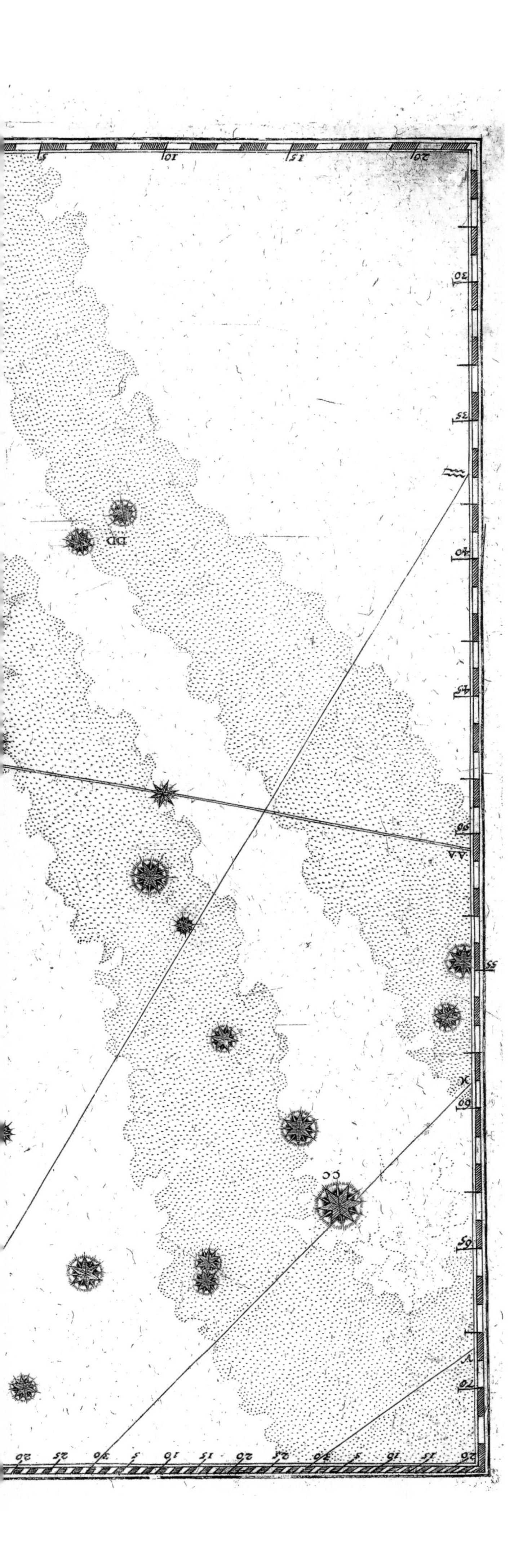

Hercules

The Hercules constellation, which is part of the Northern Hemisphere, is the fifth largest in the sky. So it comes as no surprise that it would be dedicated to such a key figure in classical mythology as the famous hero of the Twelve Labors. The demigod with prodigious strength seems to be depicted in keeping with traditional iconography, that is, armed with a club and wearing the skin of the Nemean Lion, which the hero killed during the first of his labors. However, unlike the representations of Hercules in all the other celestial maps, Bayer depicts him as seen from behind rather than frontally. In the middle of his right foot is the star indicated as Z (Z Herculis), which is rather faded. This is in fact one of the many variable stars in this constellation, which is more or less visible with the naked eye, depending on the season.

Capricornus

The Capricorn constellation, which lies in the Southern Hemisphere, is represented with the traditional ram's horns and the back portion of the body characterized by a long tail typical of sea monsters. Here too some stars appear to have been barely hatched: these are the Sigma, Pi and Omicron Capricorni, seen on the ram's snout. Bayer's decision to make these stars barely visible is most probably due to the fact that all three stars have a magnitude that varies from 5 to 6, which is the limit of visibility with the naked eye. In the middle of the creature's spiral tail is the Delta Capricorni (or Deneb Algedi *in Arabic, 'goat's tail'), the most important and luminous star in the constellation.*

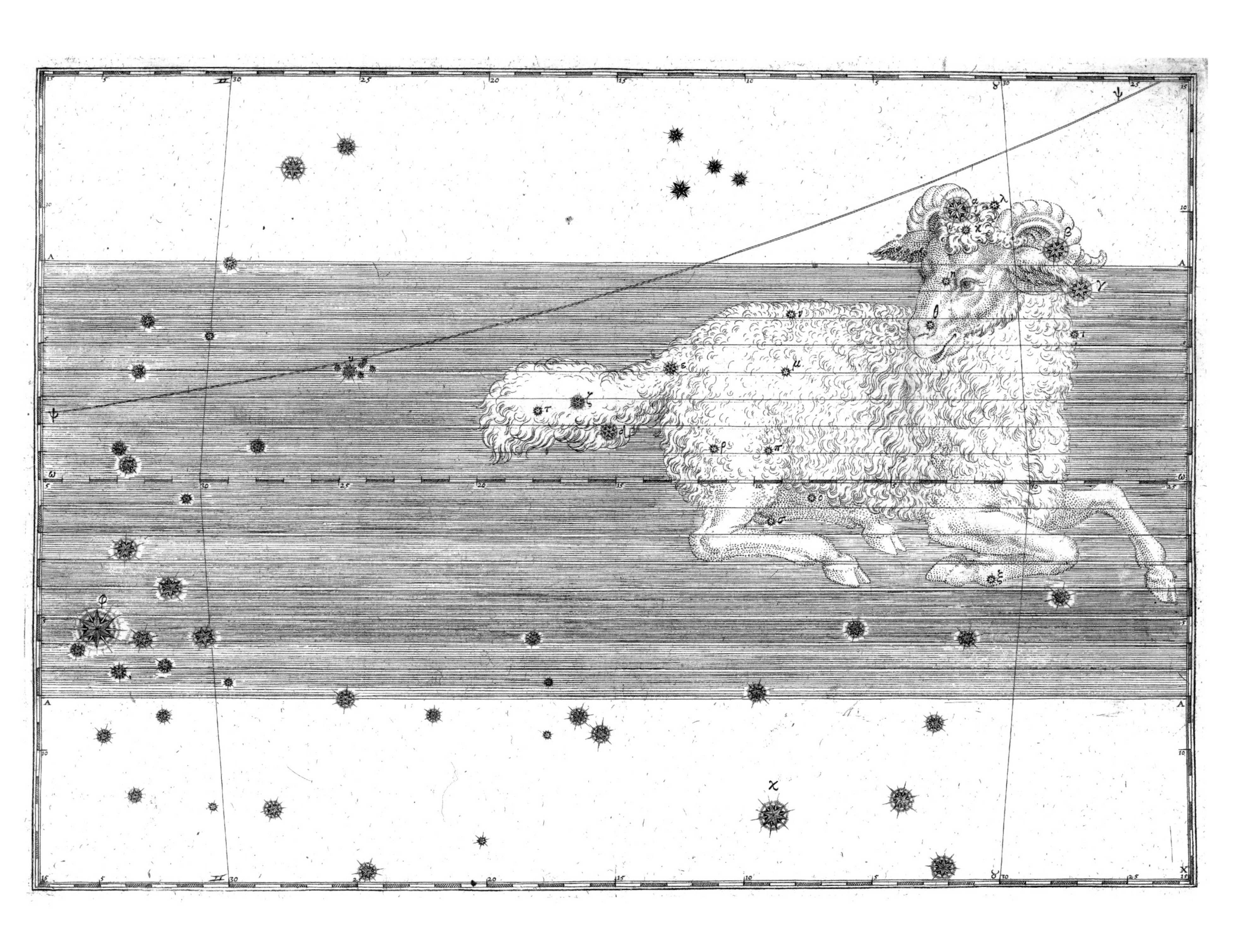

Aries

The Aries or Ram constellation represents Chrysomallus, a winged ram whose golden fleece had the power to heal wounds and for this reason was sought after by Jason. In fact he took the Argo vessel and, with Medea's help, managed to wrest it from King Aeetes. The star maps usually represent Aries without wings and in a squatting position. Bayer's plate is no exception: the animal is looking at the Taurus constellation, which lies at left (but is not seen in the plate). The three stars on the head form the characteristic asterism of this constellation, which otherwise would not be easy to distinguish. The brightest of the three is the Alpha Arietis, also called Hamal, the Arabic word for 'lamb'.

Julius Schiller

(1580-1627)

As is the case with Bayer, we have very little biographical information concerning Julius Schiller. We know only that, like his colleague, he was a lawyer at Augsburg, where he was born, presumably in 1580, and where he died in 1627. With the help of his colleague, whose passion for celestial cartographer he shared, Schiller worked on his Coelum Stellatum Christianum, *an ambitious atlas that set out to revolutionize the common perception of the cosmos. By adhering to the spirit of the Counter-Reformation, Schiller believed it was unacceptable for Catholics that the stars should be given the names of mythological divinities and creatures. Therefore, his work aimed at replacing all references to paganism with their Christian equivalents. The twelve constellations of the Zodiac were named after the twelve Apostles, those in the Northern Hemisphere after the New Testament figures, and those in the Southern Hemisphere after the protagonists in the Old Testament. On the other hand, the planets were 'reassigned' on the basis of precise individual characteristics. The Sun, the principle of light and life, was identified with Christ by reviving an ancient tradition: during the early Christian era Jesus had been identified syncretically with the* Sol Invictus *or unconquered son, the name used in the late Roman Empire for three different divinities – El-Gabal, Mitra and Sol – who often overlapped. The Emperor Constantine, who in AD 313 declared Christianity a legal religion with his Edict of Milan, was also a follower of this cult, and on 7 March 321 he established that the* Dies Solis, *the first day of the Roman week, was to be devoted to rest, so it became the* Dies Dominicus, *the day dedicated to the Lord. Furthermore, according to tradition the* Sol Invictus *was born on the winter solstice, when the days begin to become longer, thus bringing about the 'rebirth' of light. This day occurs between December 22 to 25: the Sun seems to stand still in the sky and then reverse its course, thus conquering darkness, just as Christ did through the Resurrection. The* Dies Natalis Solis Invicti, *that is, the Birthday of the Unconquered Sun, thus became the holiday of Jesus's birth or Christmas. Likewise, according to Schiller, Venus, whose appearance in the sky heralds the arrival of the rising Sun, should become St. John the Baptist, the Forerunner who announces the coming of Christ, while he renamed Mercury as Elijah, the prophet of Jesus's birth and of the Second Coming. Mars became Joshua, the 'mighty in battle' who led the people of Israel toward the Promised Land. Jupiter was rechristened Moses, the wise man who was the favorite of God and Man, and Saturn, the father of the pagan gods and the 'most distant' planet then known, became Adam, the first man. Lastly, the Moon became the Virgin Mary.*

Unlike Bayer's Uranometria, *Schiller's* Coelum Stellatum Christianum *was not greeted with much enthusiasm either by his contemporaries (except for some erudite Jesuits) or in a later period, and his proposal was soon forgotten, considered only a rather esoteric curiosity. Yet, it stands out among all the star maps for its absolute, and somewhat bold, originality.*

Constellatio XII (Sancti Hieronymi)

St. Jerome was a Father and Doctor of the Church (347-419/420) and the author of the first and most important Latin translation of the Bible, the so-called Vulgate, *which served as the basis of later versions of the Scripture up to the 20th century. In Schiller's sky the saint replaced the Auriga constellation.*

H
1
2
27
10
25
28
29
26
4
6
7
8
11
12
13
14
15
21
22
17
16
32
33
18
5
19
9
20
24
23
30
31
E

Coelum Stellatum Christianum
(1627)

Schiller's atlas basically set out to be a Christian 'revision' of Johann Bayer's *Uranometria*, which was published in 1603. Although they are bit smaller, the plates follow in precisely the same order, but the stars are still indicated with Arab numbers, thus ignoring the 'alphabetic' innovation that Bayer himself had introduced. Schiller provides a description of each star by referring to its constellation, first utilizing the pagan nomenclature and then adding the Christian names. Leaving aside the religious and ideological contents – which Schiller elaborated after consulting the Jesuit priests Johann Baptist Cysat, Paolo Guldino and Matthäus Rader – the text is interesting because it provides up-dated scientific information. The author's work is based on the studies of Tycho Brahe, while integrating data that had emerged from the recent telescopic observations made by Galileo Galilei and Simon Marius, concerning the Pleiads and the Great Andromeda Nebula (now the Andromeda Galaxy M31), which were published for the first time in this treatise. Other contemporary scholars collaborated in the creation of the *Coelum Stellatum Christianum*: besides Bayer, mention should be made of Wilhelm Schickard, an astronomer and professor of Oriental languages at the University of Tübingen; Johann Mathias Kager, who drew the constellations; and Kaspar Schecks, who checked the exact position of the stars on the copper plates that were engraved by Lucas Kilian. Schiller's work was published posthumously under the supervision of Jakob Bartsch, Kepler's son-in-law. The two plates with the 'Christian' starry heavens would be inserted by Andreas Cellarius in 1660 in his fundamental work that combined the various known theories of the cosmos, thus making them extremely famous. However, the idea of renaming the stars, constellations and planets remained only on paper: the 'traditional' pagan names had been too deeply rooted and assimilated, even in the Catholic world, to be modified in some way or other.

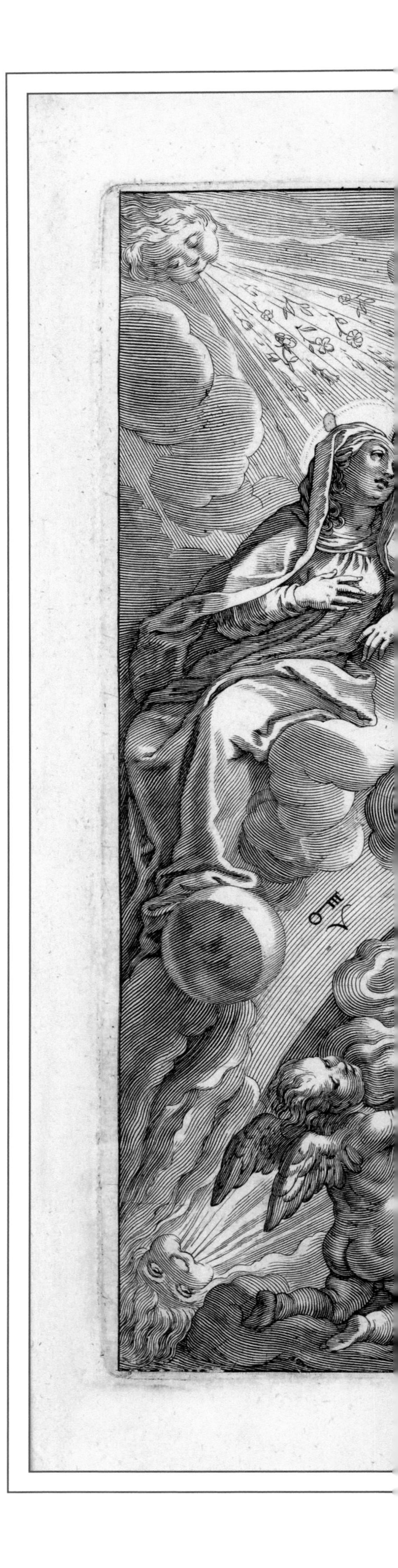

Frontispiece

Right from the frontispiece, Julius Schiller's Coelum Stellatum Christianum *reveals itself to be a work of noteworthy ideological and aesthetic depth. This engraving, which like all the others by Schiller, was executed by the German Lucas Kilian after a drawing by the painter Johann Matthias Kager, encapsulates the tenor of the revisitation of the constellations in a Christian key: the triumphant Christ (the Sun) and the protagonists of the Old and New Testament are depicted in all their glory.*

COELVM STELLATVM
CHRISTIANVM,
Ad maiorem DEI Omnipotentis, Sanctæq; Eius
tam Triumphantis, quàm Militantis ECCLESIÆ Gloriam,
Obductis Gentilium Simulachris,
EIDEM DOMINO ET CREATORI SVO,
postliminio quasi restitutum,
Humili Conatu et Voto
IVLII SCHILLERI Augustani Vindel. V.I.D.
Sociali operâ
IOANNIS BAYERI IC. Vranometriam nouam, priore
accuratiorem, locupletioremq; suppeditantis:
MATTHIÆ KAGERI, Picturam primò concinnantis.
Scalpello, quà Imagines, LVCÆ KILIANI;
quà Stellas, CASPARIS SCHECKSII:
Prælo ANDREÆ APERGERI.
AVGVSTÆ VINDELICORVM.
ANNO SALVTIS. MDCXXVII.
Cum Priuilegio Cæsareo.

C
H
N
G
A
E
D
L
M
H
I
L
K
I

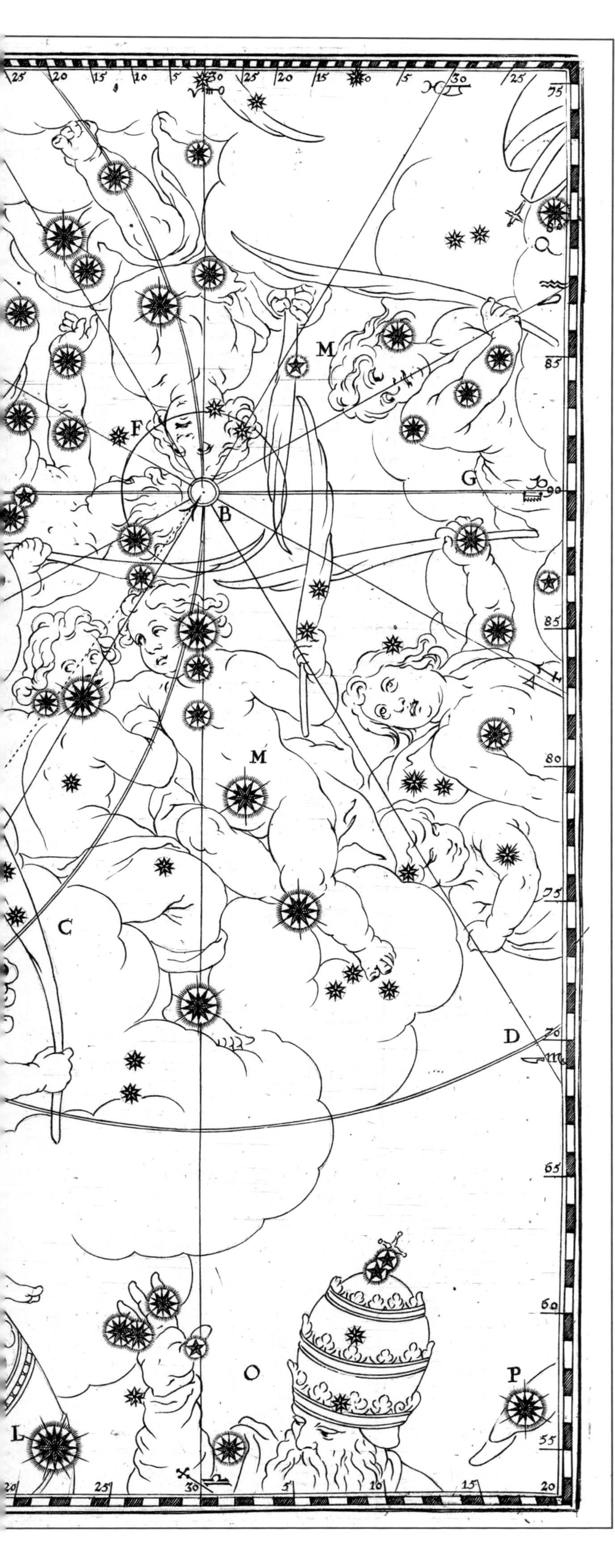

Constellatio I (Sancti Michaelis Archangeli)

St. Michael the Archangel is a central figure in Christianity since he is the leader of the angels and of God's army, he who, with his sword, cast Lucifer and the rebel angels to the earth. In Schiller's revision of the sky in a Christian key the warrior Archangel replaced the northern constellation of Ursa Minor, which is crucial because it contains the north celestial pole. The position of the Little Bear changes continuously, revolving continuously and never setting below the horizon: this is why the ancients chose its brightest star, the Polar Star (Alpha Ursae Minoris or Polaris) as an essential fixed point for journeys, by both land and sea.

CONSTELLATIO II.

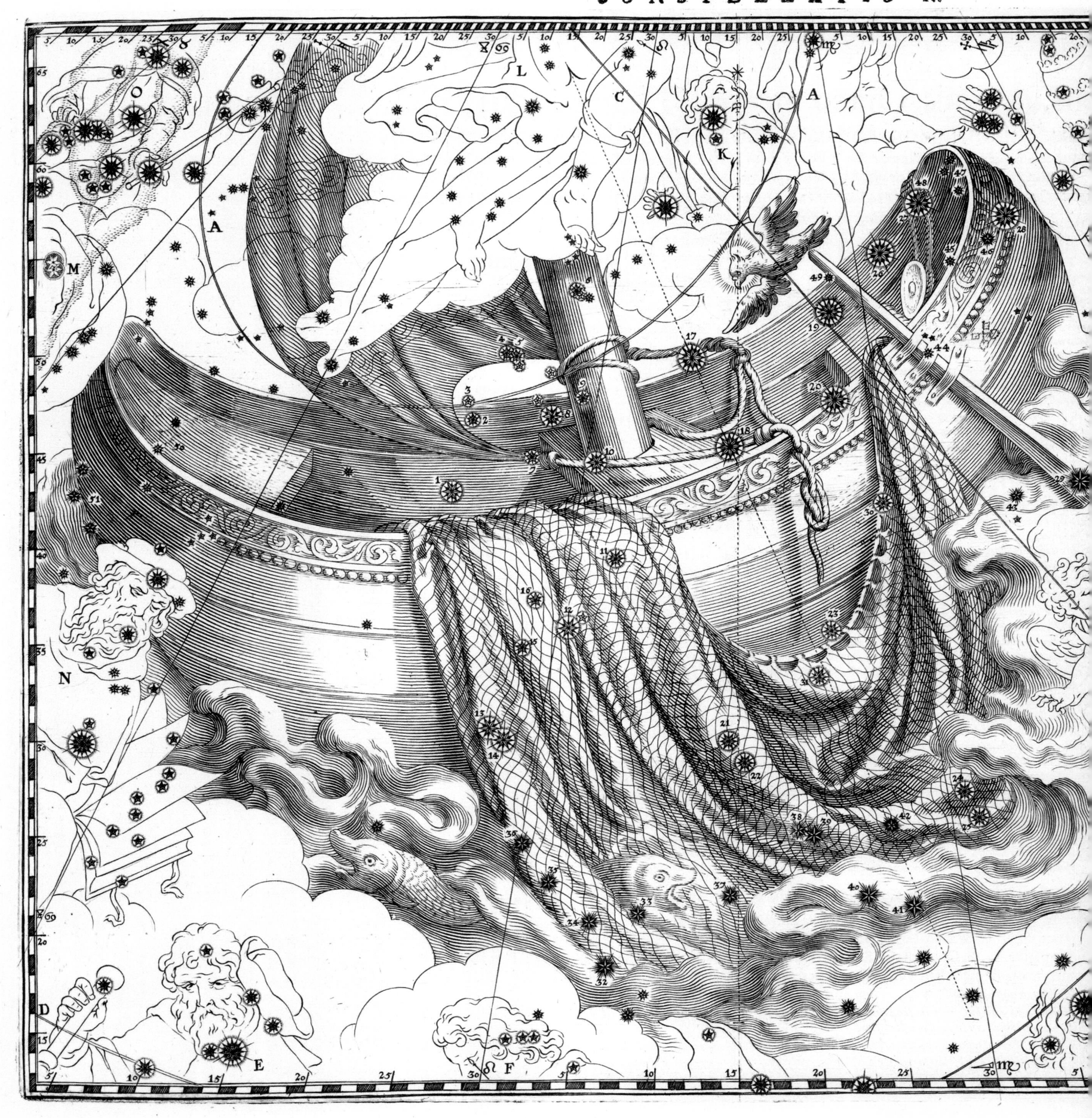

Constellatio II (Naviculae Sancti Petri)

Not far from the Ursa Minor is the Ursa Major or Great Bear, which is another northern constellation of great importance in the firmament because of its seven brightest stars, which comprise the asterism known as the Big Dipper. Like Ursa Minor, this constellation is also circumpolar: it appears to revolve without ever setting. In Greek mythology it represents the transfiguration of Callisto, a nymph of Artemis who was seduced by Zeus, who was in disguise, and was turned into a bear by the jealous Hera. Schiller called this constellation St. Peter's Boat, on which the future Apostle was casting his net when Jesus asked him to follow Him, to become a 'fisher of men': since that time it also became a metaphor for the Church.

Constellatio V (Sancti Silvestri Pontificis Maximi)

Schiller replaced the traditional name of the northern constellation Boötes (also known as Plowman) with Pope Sylvester I, a historic figure who lived between the 3rd and 4th century. According to hagiographic tradition, Sylvester I, who died in Rome on 31 December 335, is to be credited for having contributed to the (uncertain) conversion of Emperor Constantine to Christianity. Again according to tradition, Constantine, then a pagan and a persecutor of Christians, supposedly summoned Pope Sylvester to cure him of leprosy. When indeed he was cured, the emperor stopped tormenting the followers of this religion and even embraced the new faith. However, the entire story is fraught with false or legendary details that were probably invented in the 5th century to assert the pre-eminent role of the Papacy over imperial power.

Constellatio IX (S. Crucis XPI)

Thanks to the particular conformation of the Cygnus (or Swan) constellation, Schiller had no trouble seeing the facial features of Helena, mother of the Emperor Constantine. Indeed, the stars here form an asterism known as the Northern Cross, and tradition credits this saint with having discovered the True Cross of Christ, on which Jesus was crucified. A further coincidence is the fact that, according to one of the many versions of the Greek myth of Cygnus, the swan was the disguise used by Zeus in order to seduce his lover Leda: she gave birth to Helen, an extraordinarily beautiful woman who was the cause of the Trojan War. Schiller's representation has St. Helena in imperial dress, holding the True Cross and accompanied by the traditional cartouche.

CONSTELLATIO XXXV.

Bbb

Constellatio XXXV (Transitus Israel per Mare Rubrum)

In the case of this constellation Schiller displayed particularly audacious imagination. He transformed the Eridanus constellation – which is identified with the river into which Zeus hurled Phaeton, son of the sun god Helios who had stolen his father's sky chariot but lost control of it – into nothing more or less than the Red Sea which opened up so that the Hebrews escaping from Egypt and led by Moses could pass to the Promised Land. The idea probably came to Schiller from the common 'aquatic' theme of the subjects. However the rendering in the plate is quite different. The depiction of the Red Sea in itself is rather simple compared to the traditional rendering of the Eridanus constellation as a winding river. Schiller on the other hand concentrated on the 'dramatic' impact of the sudden, vertical rise of the waters, lending a great deal of emotion to the image.

CONSTELLATIO XL.

Ggg

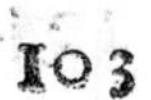

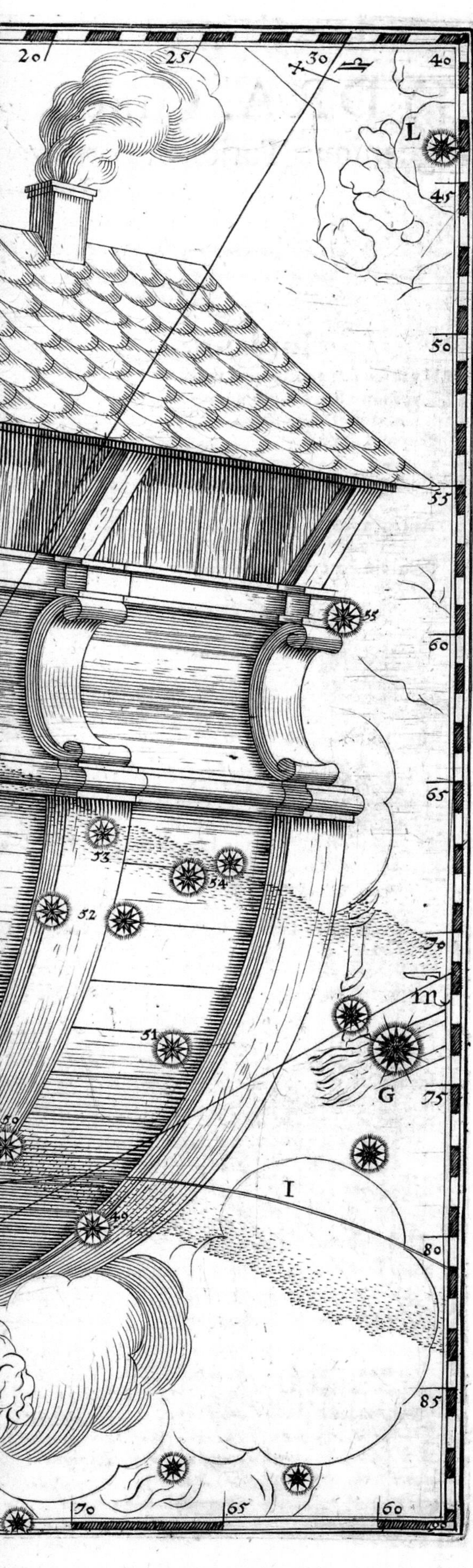

Constellatio XL (Arcae Noachi)

The Argo, the ship used by the Argonauts in their quest for the mythical Golden Fleece, was transformed by Schiller into another famous vessel – Noah's Ark. The Argo Navis constellation, which no longer exists since it was divided into three constellations, was once the largest in the sky, comprising four 'sections': Pyxis, Puppis, Carina and Vela. On the other hand, Schiller's Ark was reduced to a single entity representing the biblical ship that Noah built to save humanity and the animals from the Flood. The plate also shows Noah himself leaning out of window of the Ark to check the condition of the flooded Earth. Below left is another constellation, the Columba or Dove, naturally with an olive branch in its beak.

Constellatio XLII cum XLIII (Arcae Foederis)

In order to create the constellation of the Ark of the Alliance, this time Schiller merged two formations in the southern sky that are small and not particularly important: the Corvus (Raven), which is rather small, and the Crater, or Cup, which is not very bright and thus hardly visible. In truth, the mythological 'depth' of the two formations is also rather evanescent. The raven was sent by Apollo to fetch water in a cup, but on his way the bird saw a fig tree and decided to wait until the fruit ripened so that it could eat all it wanted. When it finally flew back to Apollo, the raven said that it had returned so late because it had been delayed by Hydra, a sea serpent. But Apollo found out the truth and transformed both animals as well as the cup into three constellations, and also condemned the raven to suffer eternal thirst, as the cup in front of the bird is out of its reach.

Constellatio XLVI et XLVII (Altaris Thymiamatis / Diadematis Salomonis)

Schiller introduced these two constellations – Altar of Incense and Solomon's Diadem – to replace Ara and Corona Australis respectively, both of which are visible in the Southern Hemisphere. According to the Greek myth, the altar of the constellation was the one in front of which the gods swore to help one another before beginning their struggle against the Titans. The Corona Australis, on the other hand, represents the wreath that the ancients used to honor the gods. In one of the versions of the myth the Corona is associated with the wreath that Dionysus dedicated to his dead mother Semele. She had been seduced by Zeus, triggering the jealousy of his consort Hera, who persuaded Semele to ask the god to appear before her in all his splendor, which, being lethal to mortals, reduced her to ashes. Both the altar and the diadem are objects mentioned in the Old Testament.

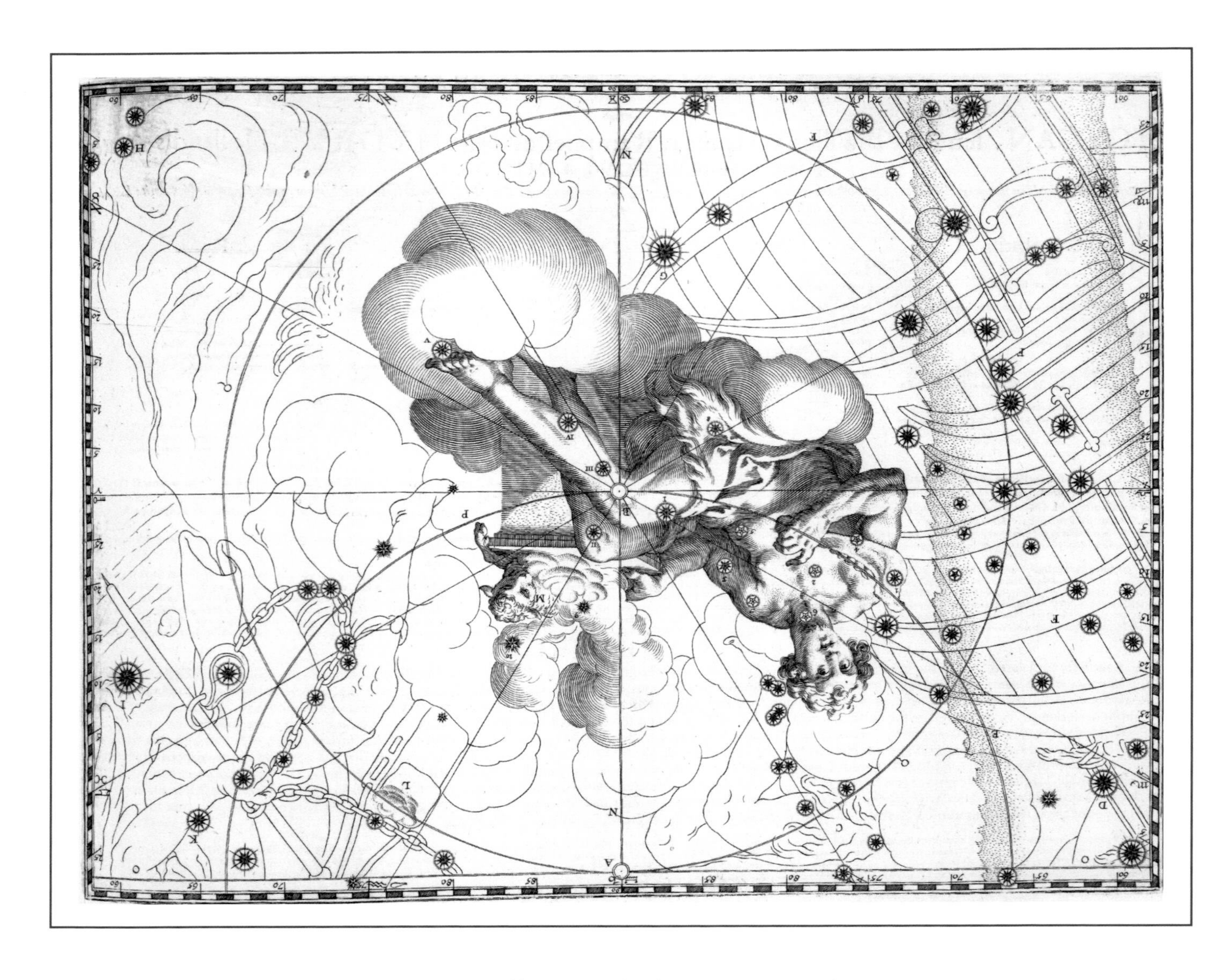

Constellatio L (Iobis Patientis)

The constellation of Job is dedicated to the patriarch and protagonist of the book in the Bible of the same name. An upright man gifted with proverbial patience, in the biblical story Job is tempted by Satan to attest to his righteousness. He loses his wealth, his seven sons and three daughters, and his health, but he bears all this suffering and stands fast in his faith. In the end God recompenses him by giving him twice as much as he had lost. Schiller put the Job constellation in place of two southern sky constellations, Pavo (Peacock) and Indus (Indian). These are two groups of 'young' stars without a solid tradition: the former was first introduced at the end of the 16th century, while Johann Bayer introduced the latter in 1603 to group together the stars that had remained 'left over' after the recent 'creations'.

Constellatio LIII (Abelis Iusti)

This constellation, dedicated to Abel, the 'just' son of Adam and Eve who was killed by his 'evil' brother Cain, was also conceived by Schiller by uniting two southern celestial constellations that had been created recently and therefore lacked important historical and mythological tradition. They both drew inspiration from fish discovered in the New World. The first is Dorado, introduced in the likeness of a dolphinfish or dorado, while the second is Volans, the Flying Fish, a tropical fish able to jump out of the water and 'fly' briefly by using two fins as 'wings'. These two 'ichthyological' constellations had been conceived in the 16th century (together with ten others of differing inspiration, making a total of twelve) by Petrus Plancius and taken up later in celestial atlases, first of all in Bayer's. However, the German astronomer had preferred to represent Dorado as a swordfish, giving it the name of Xiphias.

Andreas Cellarius

(1596-1665)

Very little about the life of Andreas Cellarius has been handed down to us, and his existence is still wrapped in 'mystery' despite all the meticulous research made in this direction. On the title page of his most important work, Harmonia Macrocosmica, *published in 1660, which brought lasting fame to Cellarius, the author describes himself only as "rector of the Latin school in Hoorn, in Northern Hollands". Archive documents in Amsterdam and other cities in Holland have added very little to this already very meager piece of information: only the year of birth, 1596, and the birthplace, Neuhausen, a German town not far from Worms, in Rhineland-Palatinate. Other documents reveal that his father, who was also christened Andreas, was a Protestant preacher and that in 1599 he and his family moved to Heidelberg, in Baden-Württemberg, where Cellarius attended the local university and then most probably made a long trip to Poland; but we do not know why he went there, or who, if anyone, had summoned him. What is certain is that years later he made the most of what he had learned during this journey in a long treatise containing a description of the institutions and geographic features of the Kingdom of Poland and Duchy of Lithuania. We can surmise that on this occasion he began to become interested in cartography, an interest that later led him to broaden his view from the limited horizons of the Earth to the infinite expanse of the sky.*

Nothing is known about Cellarius's pursuits up to 1625, the year his name suddenly appeared in the matrimonial register of the city of Amsterdam together with that of his bride, Catharina Eltmans. The following year the couple had a son, who was also christened with the name of Andreas, and two years later they had a daughter, Catharina. After the marriage his life was most probably quite uneventful, his time spent between the family – Andreas and the little Catharina soon had two more siblings, Johannes and Joris – and his teaching in the Latin schools, first in the Old Town of Amsterdam and then, in 1630, in The Hague, where he obtained a post in the former St. Agnes Monastery, on the Zuilingstraat. Lastly, in 1637 Cellarius moved to Hoorn, where he was appointed rector of the former monastery of St. Cecilia and where he could devote his energy to study and writing in peace and quiet.

In this city, which was then the capital of the historic region of West-Friesland, he had all his works published, beginning with the first one we know of, Architectura Militaris *(1645), a long treatise in which Cellarius examines the various types of fortifications and defense systems that would be efficacious against sieges. After publishing the above-mentioned description of Poland – which came out in 1652 with the title* Regni Poloniae, Magnique ducatus Lituaniae: Omniumque regionum juri Polonico Subjectorum: Novissima descriptio *and was reprinted in 1659 – Cellarius began to work on his* Harmonia Macrocosmica, *an illustrated compendium of the existing theories of the cosmos, which he intended to be the first part of a wide-ranging work,* Atlas Novus, *which was never published.*

The importance of this work does not lie in its scientific formulation – which in fact was criticized by the intellectuals of the time, so much so that it had to be defended by the Jesuit Athanasius Kircher to prevent it from being placed on the Vatican Index of Prohibited Books – but in the stunning beauty of its plates, executed by the Dutch engraver and mathematician Jan van Loon, some of which were colored by hand. Cellarius died in Hoorn sometime between February and March 1665. His burial site is unknown. Besides his masterpiece – which the astronomy historian Robert van Gen calls the finest celestial atlas ever produced – he is also immortalized by an asteroid named after him in 2008: 12618 Cellarius.

Portrait of Tycho Brahe

The Danish astronomer Tycho Brahe, whose theories were cited by Cellarius, in a detail of the plate dedicated to him *(see pages 66-67)*.

VIR GO
LE O
30
20
10

Harmonia Macrocosmica
(1660)

This work, published in Hoorn in 1660 by the Dutch publisher and engraver Johannes Janssonius (1588-1664) and reprinted the following year, appeared as the summa, written in Latin and about 200 pages long, of the various cosmological systems that were known at that time. These range from the Ptolemaic geocentric system to Copernicus's model of the solar system and to the one proposed by the Danish astronomer Tycho Brahe, which was a sort of compromise between the first two, ending with the reinterpretation of the constellations in a biblical key that Julius Schiller had elaborated only a few decades earlier. Cellarius does not evince a preference for any of these approaches but merely limits his contribution to illustrating them through 29 spectacular large-format double-folio copper plates. These were originally printed in black and white and at a later stage some volumes were colored by hand, thus creating a series of unique works.

The *Harmonia Macrocosmica* was preceded, the previous year, by the publication of two plates representing the sky as viewed by the ancients, divided into the Northern and Southern Hemispheres: both these illustrations appeared in the Atlas with the numbers 24 and 27 respectively. Cellarius had originally intended his *Harmonia Macrocosmica* to be a larger atlas divided into two sections: the first containing the compendium of the theories put forward in the past, while the second would be a treatise dealing with the theories of his time. However, the latter section was never written, so that only the illustrated historical synthesis remains.

Frontispiece

The frontispiece of Andreas Cellarius's Harmonia Macrocosmica seu Atlante universalis et novus, totius universi creati cosmographiam generalem, et novam exhibens *is nothing more or less than a true work of art. It represents Urania, the Muse of Astronomy, surrounded by the principal ancient and recent astronomers as well as globes, quadrants and armillary spheres. Ptolemy is behind Urania, while Copernicus is at her left and Tycho Brahe at her right.*

ATLAS COELESTIS;
seu
HARMONIA
MACROCOSMICA.

Planisphaerium Ptolemaicum

The first plate of the work is titled Planisphaerium Ptolemaicum, sive Machina Orbium Mundi ex Hypothesi Ptolemaica in Plano Disposita *and represents Ptolemy's planisphere with the 'mechanisms', that is, the movements of the celestial orbits he describes. Obviously, Earth is placed in the middle of the Solar System, and seven celestial bodies revolve around it, from the Moon to Jupiter (including the Sun, which is in the fourth position). Their orbits are represented as concentric circles, each of which has an image of the celestial body steering a chariot. The Earth seems to be surrounded by a mass of nimbi and a fiery crown that envelops the 'sub-lunar world', including the four elements: fire, air, water and earth.*

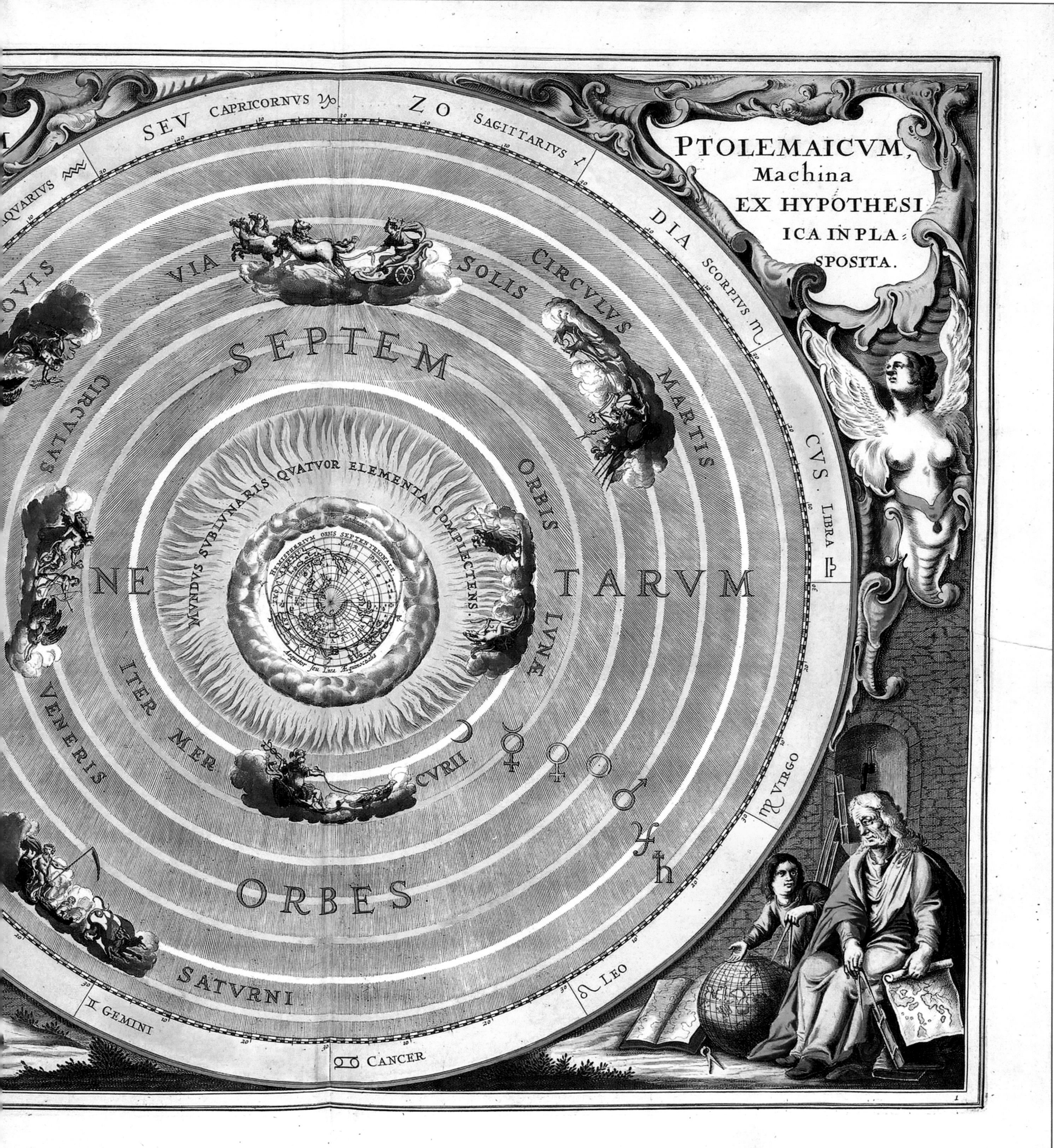
PTOLEMAICVM,
Machina
EX HYPOTHESI
ICA IN PLA:
SPOSITA.
SEV
CAPRICORNVS
ZO
SAGITTARIVS
AQVARIVS
DIA
SCORPIVS
CVS
LIBRA
VIRGO
LEO
CANCER
GEMINI
VIA SOLIS
CIRCVLVS MARTIS
SEPTEM
CIRCVLVS
ORBIS LVNÆ
NE
TARVM
MVNDVS SVBLVNARIS QVATVOR ELEMENTA COMPLECTENS.
ITER MERCVRII
VENERIS
ORBES
SATVRNI
Æquator seu Linea Æquinoctialis

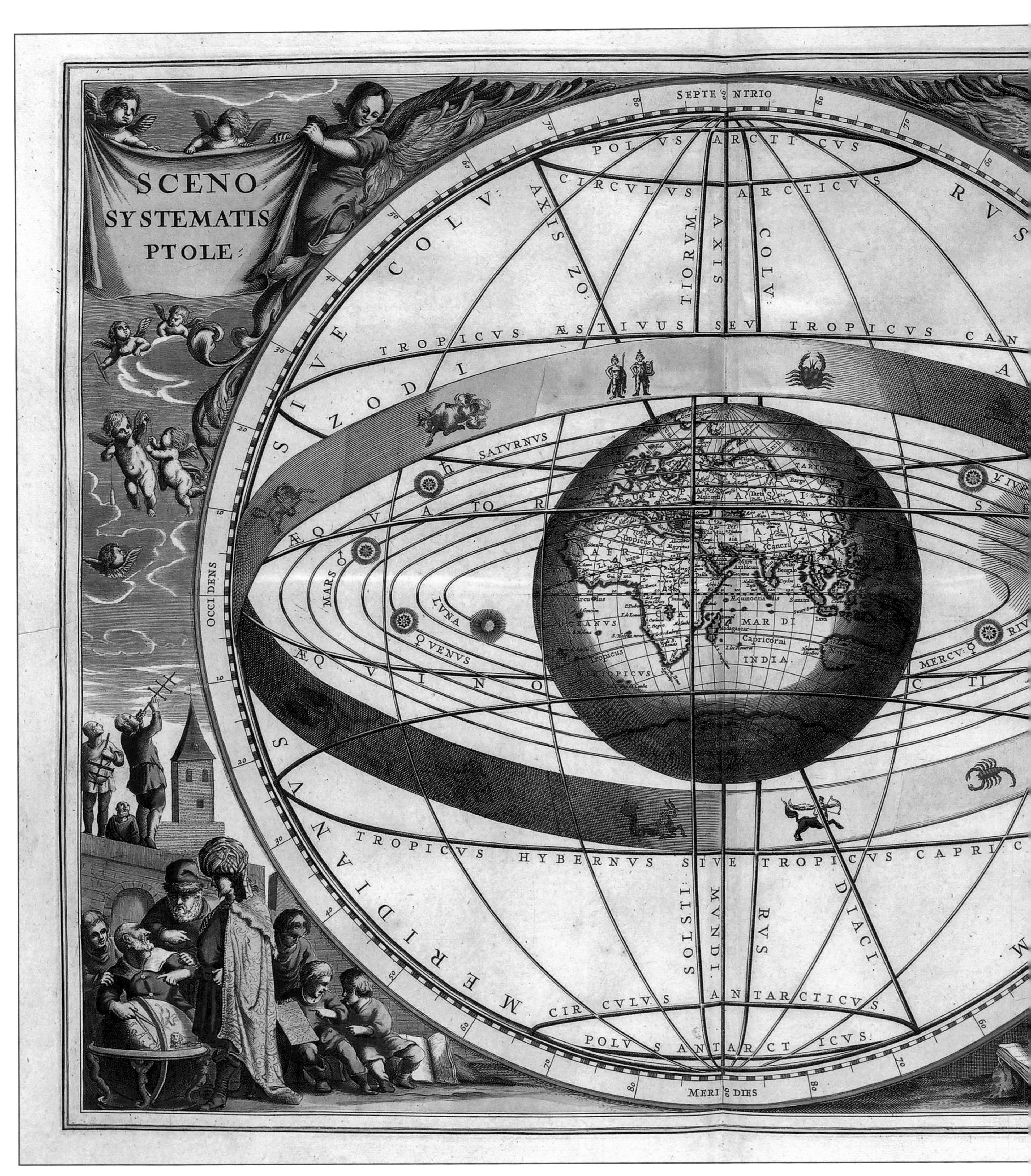
SCENO
SYSTEMATIS
PTOLE:
SEPTE NTRIO
POLVS ARCTICVS
CIRCVLVS ARCTICVS
SIVE COLV:
AXIS ZO:
TIORVM.
AXIS
COLV:
TROPICVS ÆSTIVUS SEV TROPICVS CAN
ZODI
SATVRNVS
ÆQVATOR
MARS
LVNA
VENVS
MERCV:
AEQVINOCTI
OCCIDENS
EVROPA
ASIA
AFR
MAR DI
Capricorni
INDIA
TROPICVS HYBERNVS SIVE TROPICVS CAPRIC
MERIDIANVS
SOLSTI:
MVNDI
RVS
DIACI
CIRCVLVS ANTARCTICVS.
POLVS ANTARCTICVS.
MERI DIES

Scenographia Systematis Mundani Ptolemaici

In this plate, the second in the atlas, Cellarius summarizes the cosmographic theories of Ptolemy in a 'scenography'. Earth is placed in the middle of the universe, with the planets (including the Sun) and the constellations of the Zodiac revolving around it. The author chose to depict Earth from a 'euro-centric' perspective: besides the Old World, one can see Africa and Asia, and even part of Antarctica. At right is a partial view of Nova Hollandia (New Holland), which was explored for the first time by the Dutch in 1606 and was named by the Dutch navigator and explorer Abel Tasman in 1644. However, at that time this name was used to embrace the entire south of Earth (or Australis), which would later take on the name of Australia.

Orbium Planetarum Terram Complectentium Scenographia

Plate number 3, once again inspired by the Ptolemaic system, represents in a 'scenographic' manner the orbits of the planets around Earth. The plate is quite striking in that it resembles a complex armillary sphere. Thanks to the perspective, one can appreciate the movement of the Sun, the Moon and the planets, also in relation to the Zodiac belt. In typical baroque style, both upper corners of the plate contain two cherubs holding up a cloth announcing the subject dealt with. On the lower corners the same cherubs illustrate with two diagrams the opposing theories proposed by Ptolemy (left) and Tycho Brahe (right).

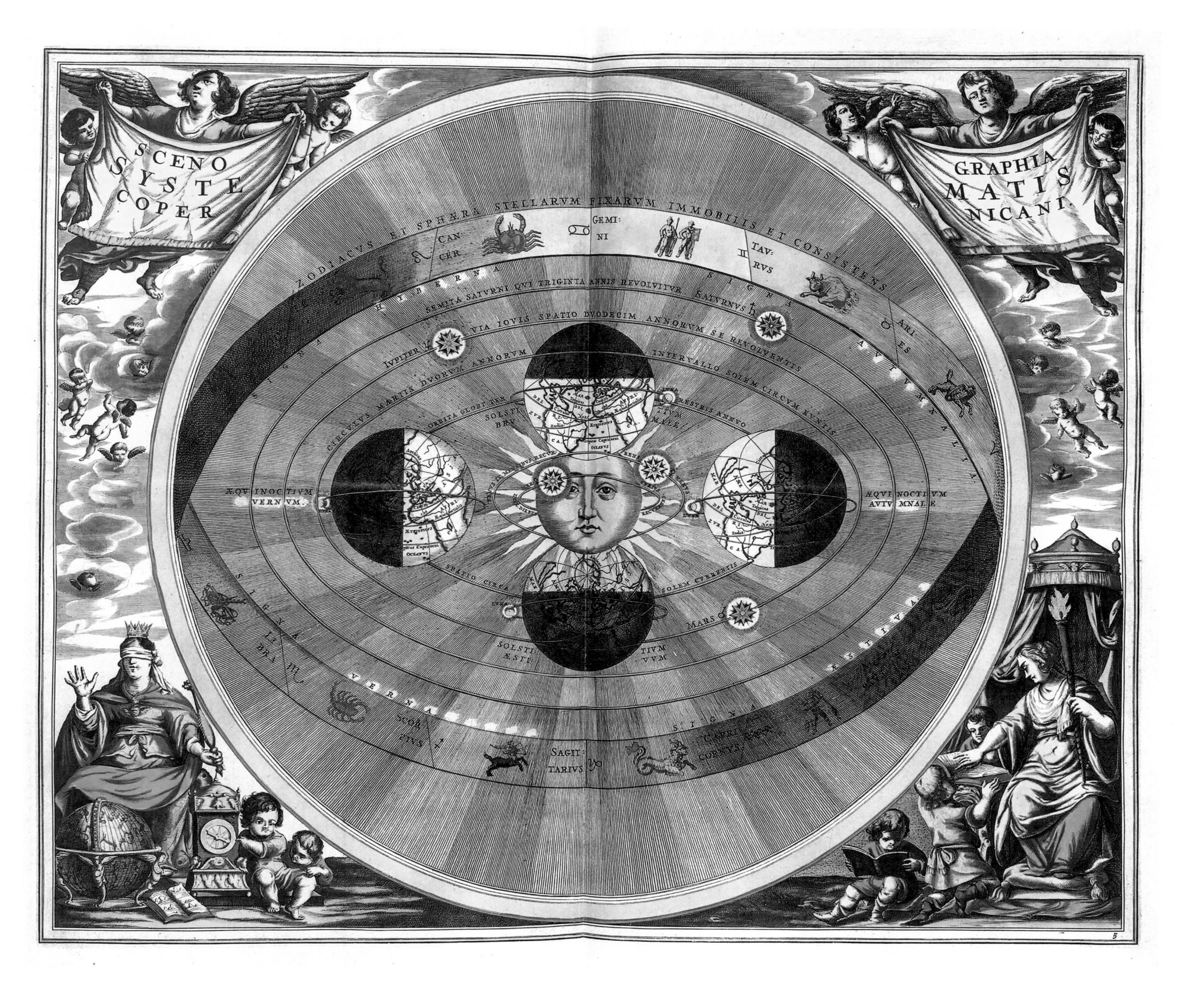

Scenographia Systematis Copernicani

This map, the fifth of the series, is an intriguing scenographic representation of the Copernican world system. Unlike the Ptolemaic system, which placed Earth in the center of the universe, here the Sun is its hub. The four images of Earth indicate the positions of our planet with respect to the four seasons, with the hemisphere opposite the Sun depicted in the dark. These are followed by the orbits of the individual planets (errant bodies), which according to ancient and medieval thought moved around Earth like so many concentric spheres. The last of these is the sphere of the fixed stars. On the lower corners of the scene are allegories of Fortune (left) and perhaps of Knowledge (right), while the usual winged cherubs decorate the four corners.

Planisphaerium Copernicanum

The fourth plate in the Atlas, the title of which is Planisphaerium Copernicanum Sive Systema Universi Totius Creati ex Hypothesi Copernicana in Plano Exhibitum, *presents the planisphere (that is, the universe) according to Copernicus. The system is heliocentric: the Sun is depicted in the middle, with the planets revolving around it in perfectly concentric orbits. Earth is depicted in detail, as if it were viewed from above (from the North Pole, to be precise) and with the Moon revolving around the Earth. Besides the aesthetic beauty of this plate (which is especially enhanced in the hand-colored versions), note the presence around Jupiter of its four satellites, which were discovered by Galileo Galilei in 1610 and which he called 'Astri Medicei' in his* Sidereus Nuncius.

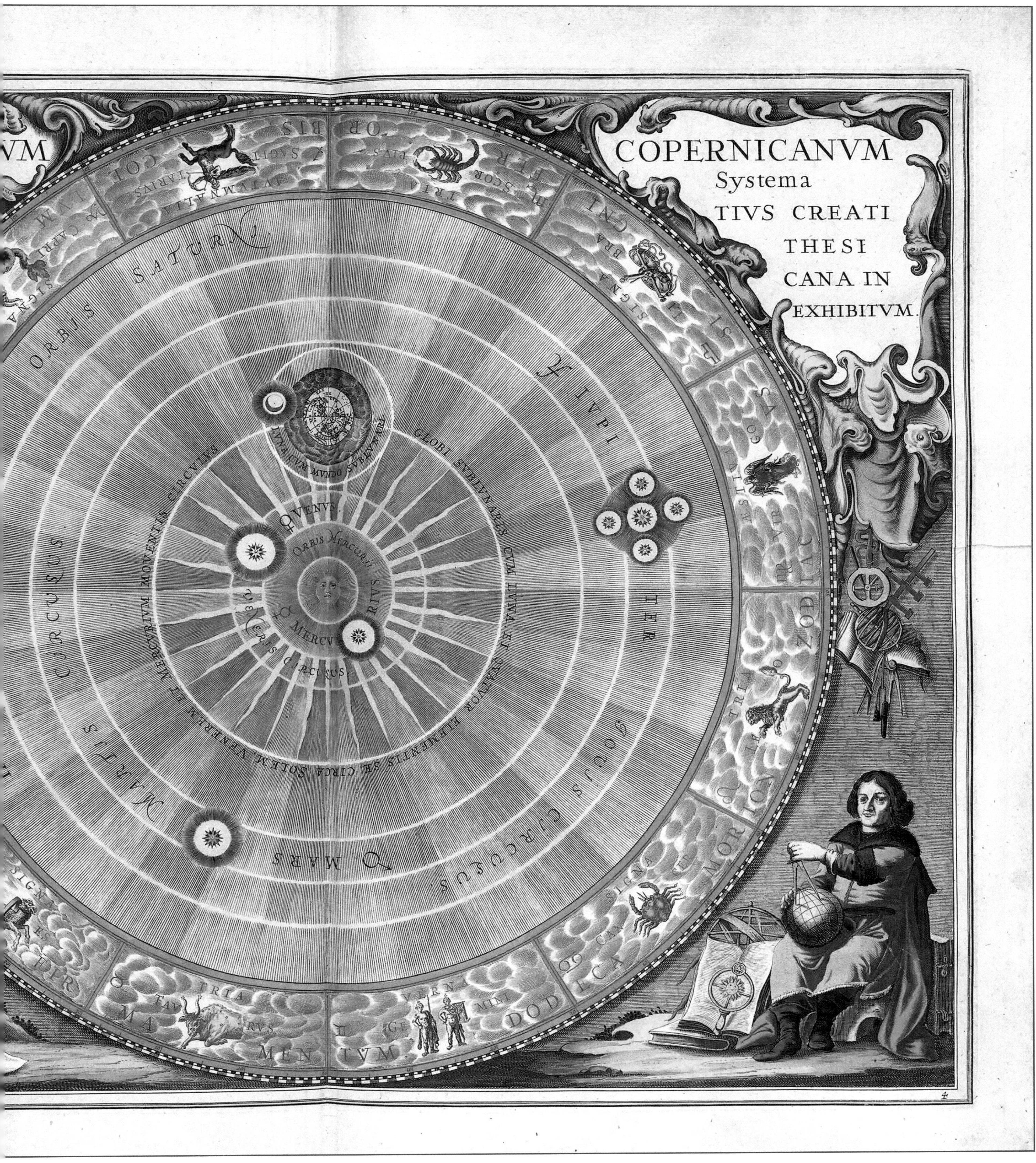

COPERNICANVM
Systema
TIVS CREATI
THESI
CANA IN
EXHIBITVM.
ORBIS SATURNI
IUPI TER
MARTIS CIRCUUS
MARS
VENUS
ORBIS Mercurii
MERCURIUS
LUNA CUM MUNDO SUBLUNARI
GLOBI SUBLUNARIS CUM LUNA EI QUATUOR ELEMENTIS SE CIRCA SOLEM VENEREM ET MERCURIUM MOVENTIS CIRCULUS
ZODIAC
4

Planisphaerium Braheum

The Planisphaerium Braheum, Sive Structura Mundi Totius, ex Hypothesi Tychonis Brahei in Plano Delineata *is the sixth plate in the series and represents the planisphere of Tycho Brahe. The Danish astronomer (1546-1601) made an in-depth study of the Ptolemaic and Copernican systems of interpretation of the cosmos, arriving at the elaboration of a third theory that lay between the other two, the so-called Tychonic system: Earth is immobile in a central position, the Sun revolves around it, while the other planets – except for the Moon, which also revolves around Earth – move around the Sun itself. Again, also revolving around Jupiter are its four satellites. Below right is a portrait of Brahe in his famous observatory and research center of Uraniborg on the island of Hven, a gift of Frederick II of Denmark and Norway.*

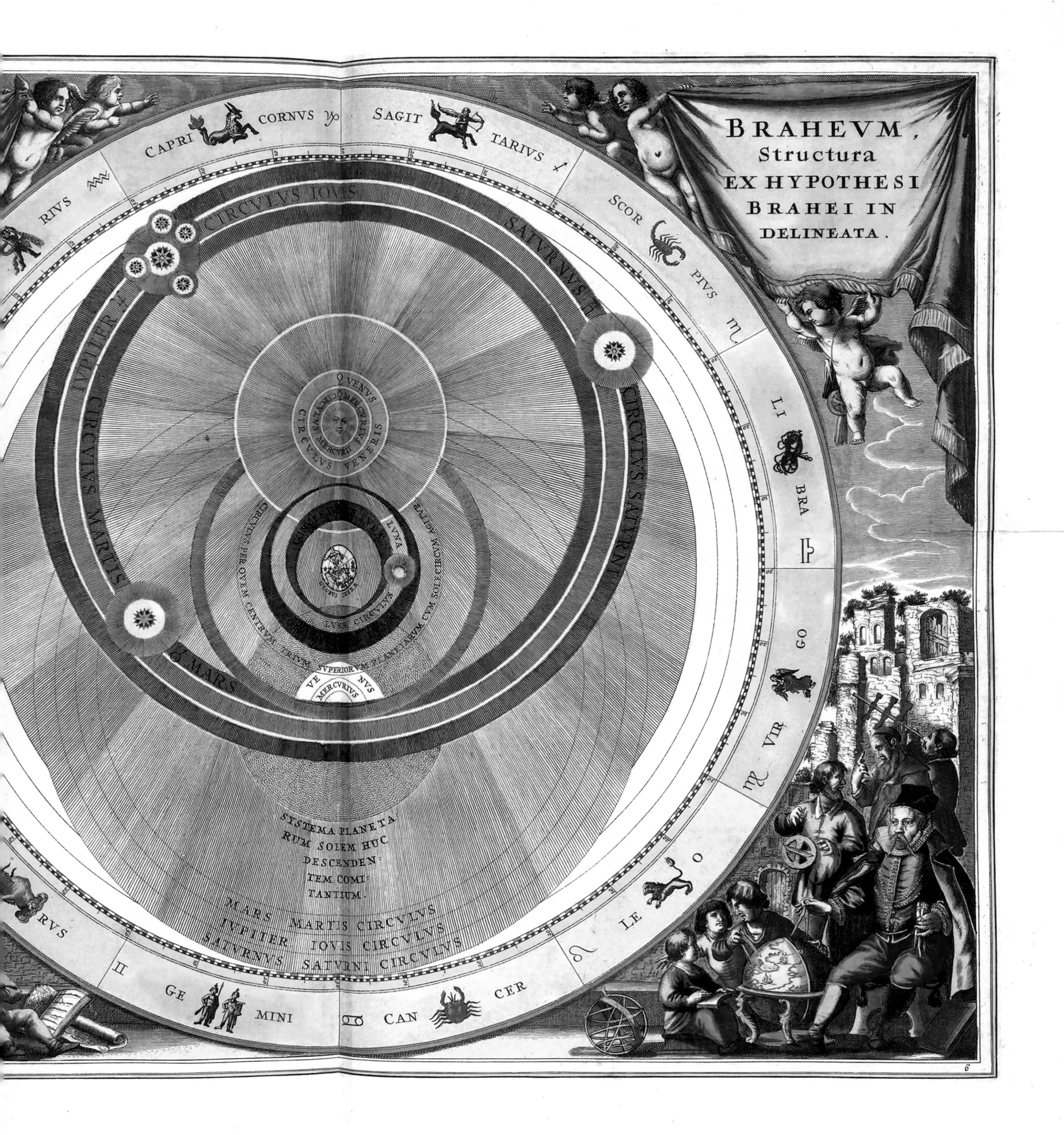
BRAHEVM,
Structura
EX HYPOTHESI
BRAHEI IN
DELINEATA.
CAPRI CORNVS
SAGIT TARIVS
SCOR PIVS
LI BRA
VIR GO
LE O
CAN CER
GE MINI
RVS
RIVS
CIRCVLVS IOVIS
SATVRNVS
CIRCVLVS SATVRNI
IVPITER
CIRCVLVS MARTIS
MARS
VENVS
MERCVRIVS
CIRCVLVS MERCVRII
CIRCVLVS VENERIS
LVNA
LVNÆ CIRCVLVS
CIRCVLVS PER QVEM CENTRVM TRIVM SVPERIORVM PLANETARVM CVM SOLE CIRCVMAGITVR
VE NVS
MERCVRIVS
SYSTEMA PLANETA
RVM SOLEM HVC
DESCENDEN
TEM COMI
TANTIVM.
MARS MARTIS CIRCVLVS
IVPITER IOVIS CIRCVLVS
SATVRNVS SATVRNI CIRCVLVS
6

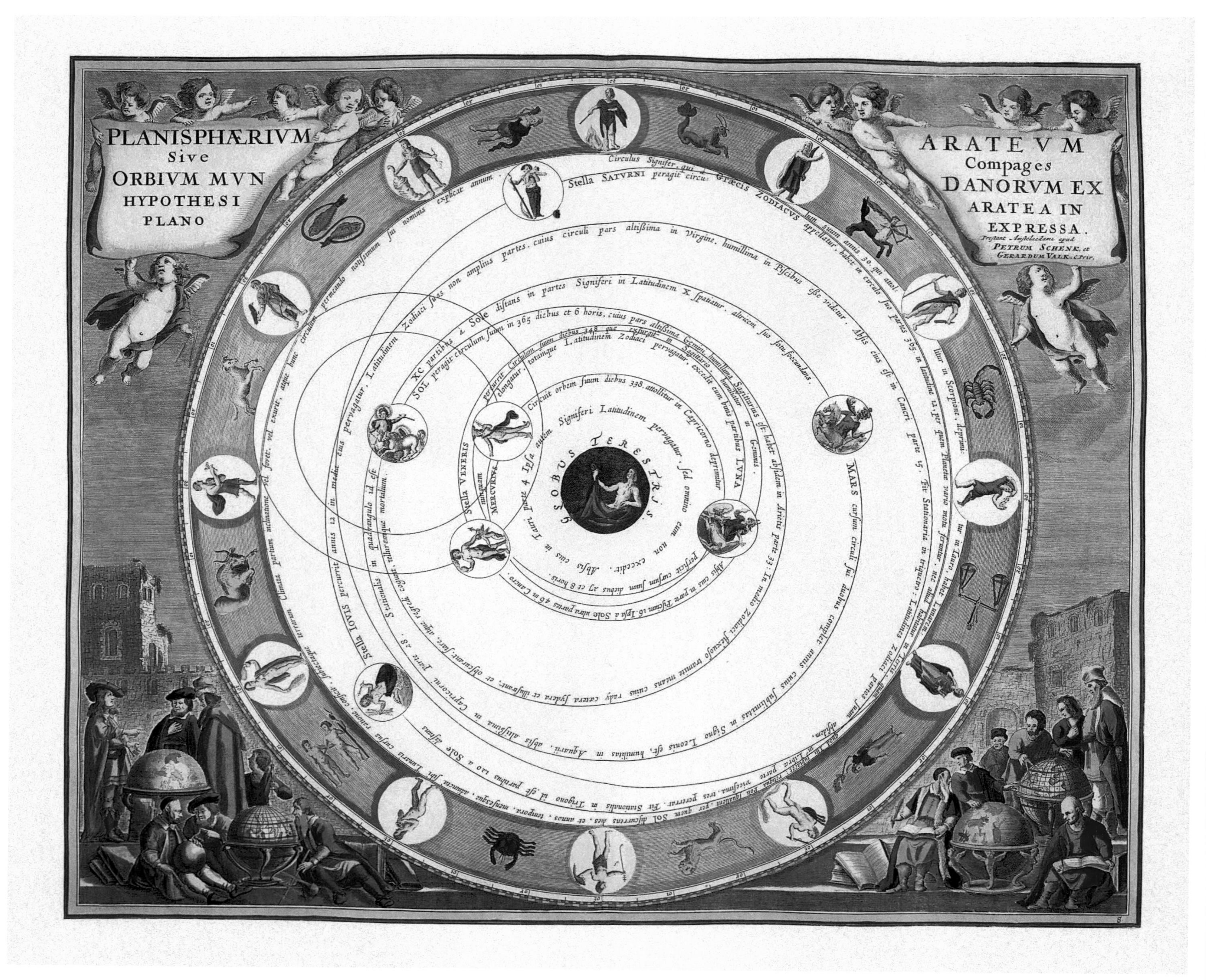

Planisphaerium Arateum

Plate number eight of the Harmonia Macrocosmica, *which is titled* Planisphaerium Arateum Sive Compages Orbium Mundanorum ex Hypothesi Aratea in Plano Expressa, *represents the planisphere according to Aratus of Soli (c. 315–c. 240 BC). The ancient Greek poet dealt with astronomy in the first part of his* Phaenomena (Appearances), *a didactic poem written in hexameters. Taking up the theories formulated by the astronomer Eudoxus of Cnidus – which, naturally, are geocentric – Aratus described the constellations, their rising and setting, as well as the circles that divide the celestial sphere. This was known in the Western world during the Middle Ages thanks to the numerous Latin translations, so that the* Planisphaerium *had an enormous influence on astronomy before the age of scientific discoveries.*

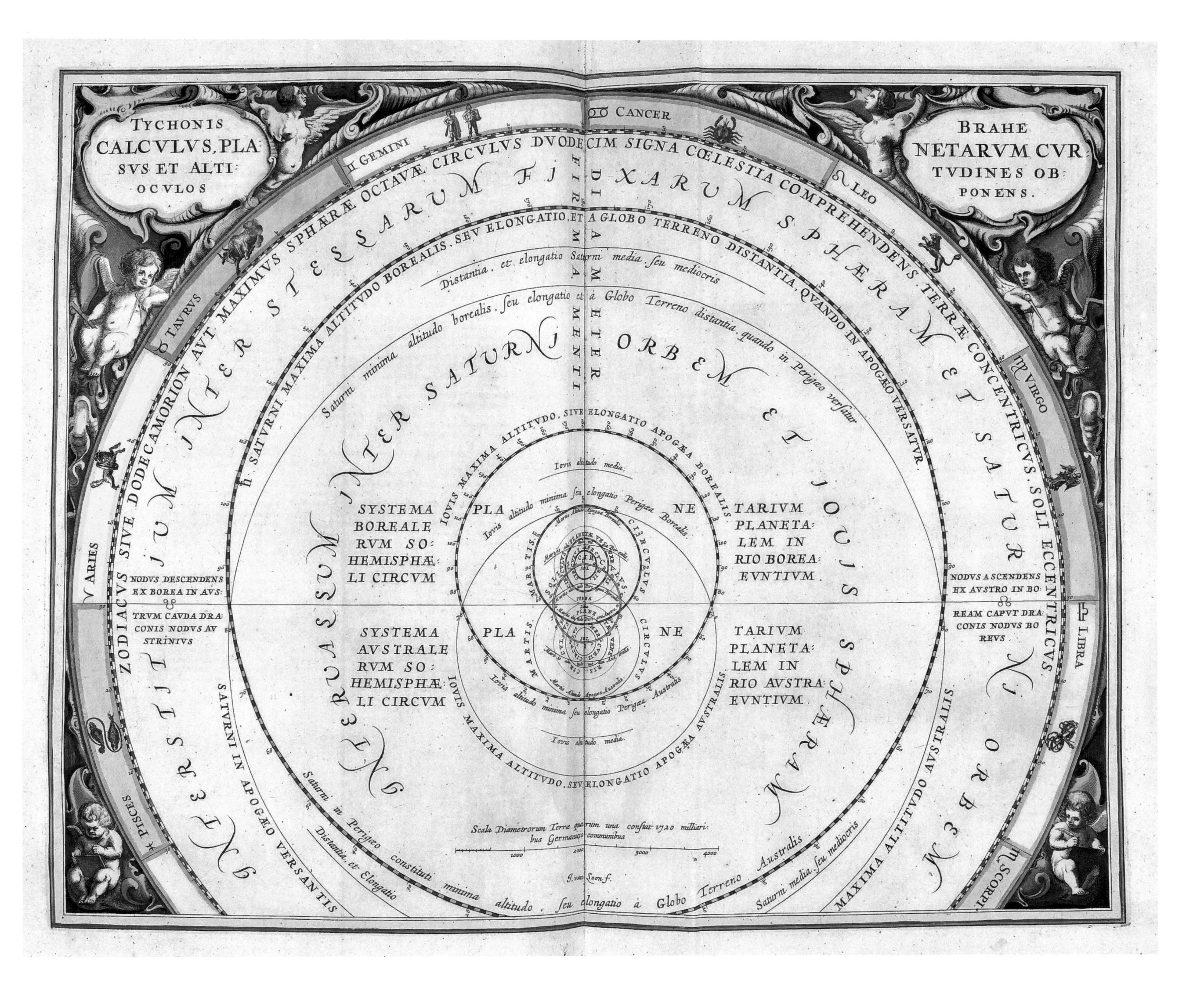

Tychonis Brahe Calculus

*In this plate, the ninth, Cellarius proposes a synthesis of the method formulated by Tycho Brahe to calculate the course and distances of the planets (*Tychonis Brahe Calculus, Planetarum Cursus et Altitudines ob Oculos Ponens*). Brahe was convinced that Earth, being composed of the four elements (water, land, air and fire), was too slow and heavy to be constantly moving, as is also demonstrated by the fact that on Earth an object cannot move by itself but must be moved. Vice versa, according to Aristotelian physics the heavens, being made up of light and impalpable air, were always moving. Hence the need to calculate the movement of the planets according to different principles, which are summarized by Cellarius in this plate.*

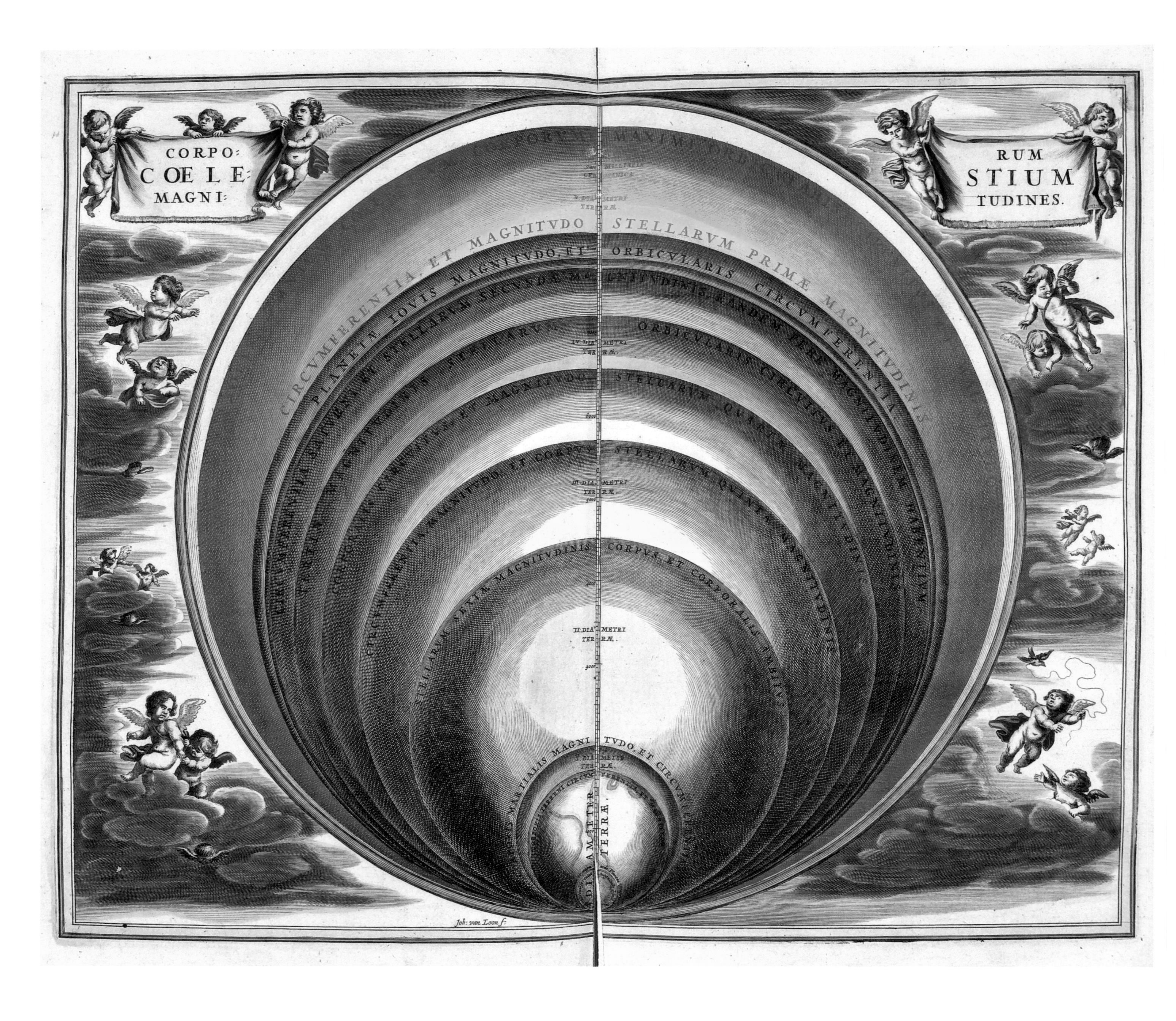

Corporum Coelestium Magnitudines

*Plate ten is given over to the dimensions (*magnitudines*, in Latin) of the celestial bodies compared to our planet. This calculation was made possible not only by the visual comparison of the sizes of the stars, but also through the vertical ruler in the middle, the spaced markings of which measure the diameter of Earth. The celestial bodies whose size is provided by Cellarius are both stars and planets, and they are all arranged in progressive order, from the smallest (*magnitudine sexta*) to the largest (*magnitudine prima*). The Sun, which is the largest star of all, is described as 'beyond measurement'. A curiosity: in some editions of* Harmonia Macrocosmica *the various continents of Earth were designed as well.*

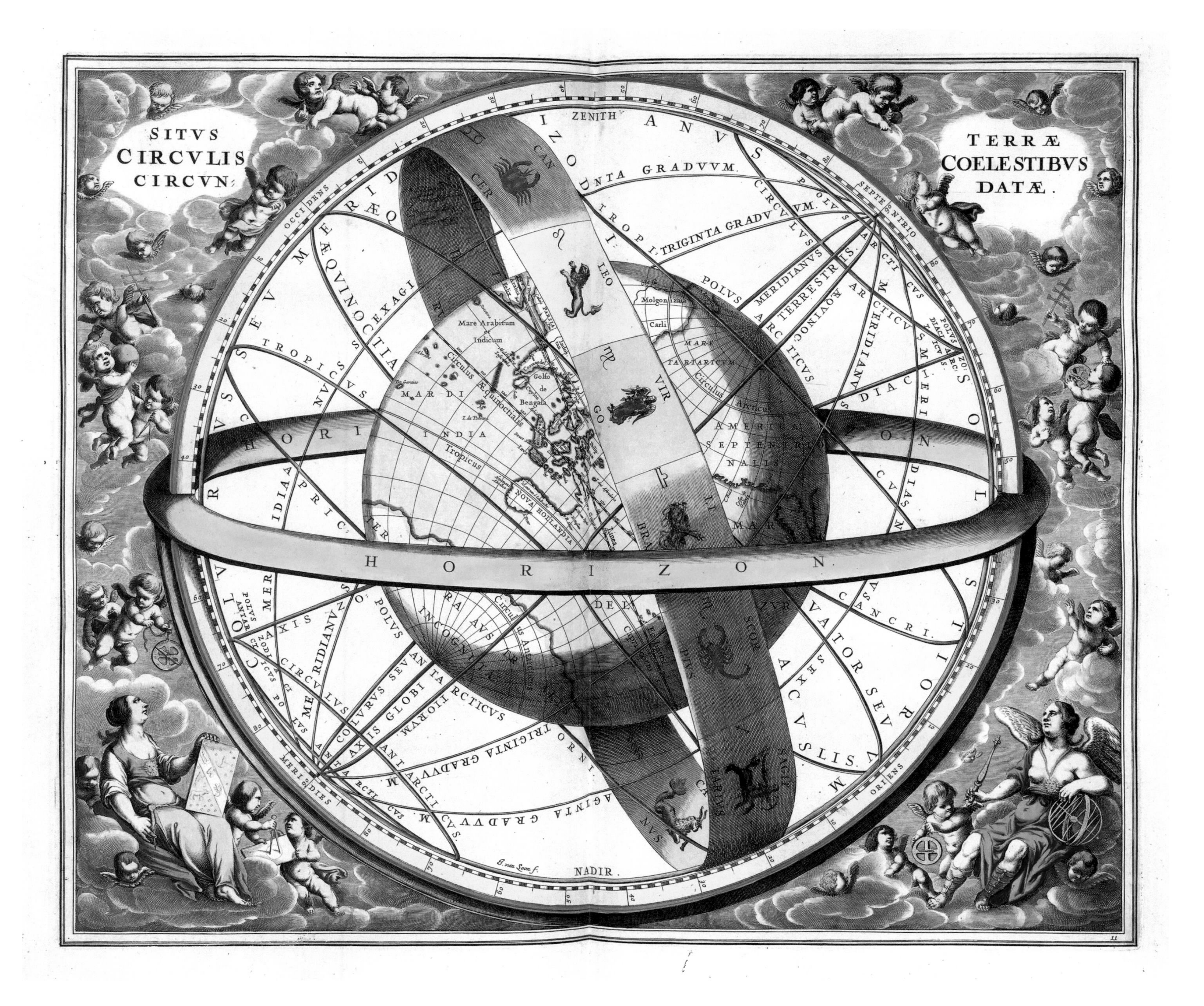

Situs Terrae Circulis Coelestibus Circundatae

This plate, the eleventh, features the Earth – placed in the middle of the sky in keeping with Ptolemy's theory –surrounded by the horizon and the Zodiac with its constellations in keeping with the apparent movement of the Sun. Note, on the globe, the indefinite (or open-ended) confines of Nova Hollandia, or Australia. In fact, exploration of this continent had begun a short time before publication of Cellarius' work and was still in progress. California, on the other hand, in both this and other plates, is erroneously depicted as an island. The usual baroque decoration appears on the corners, with cherubs and two female figures in the lower corners who are holding astronomical instruments. The woman at right may be Urania, the Muse of astronomy.

Haemisphaeria Sphaerarum Rectae et Obliquae

Plate 12, the complete (and long) title of which is Haemisphaeria Sphaerarum Rectae et Obliquae Utriusque Motus et Longitudines Tam Coelestes Quam Terrestres ac Stellarum Affectiones Mons[t]rantia, *shows the movement of both the celestial and terrestrial spheres, with their orbits and epicycles. Among the meridians are the equinoctial colure (the meridian of the celestial sphere that passes through the celestial poles and the spring and autumn equinoxes), the solstitial colure (the meridian that passes through the celestial poles and the two solstices), the celestial equator and the ecliptic.*

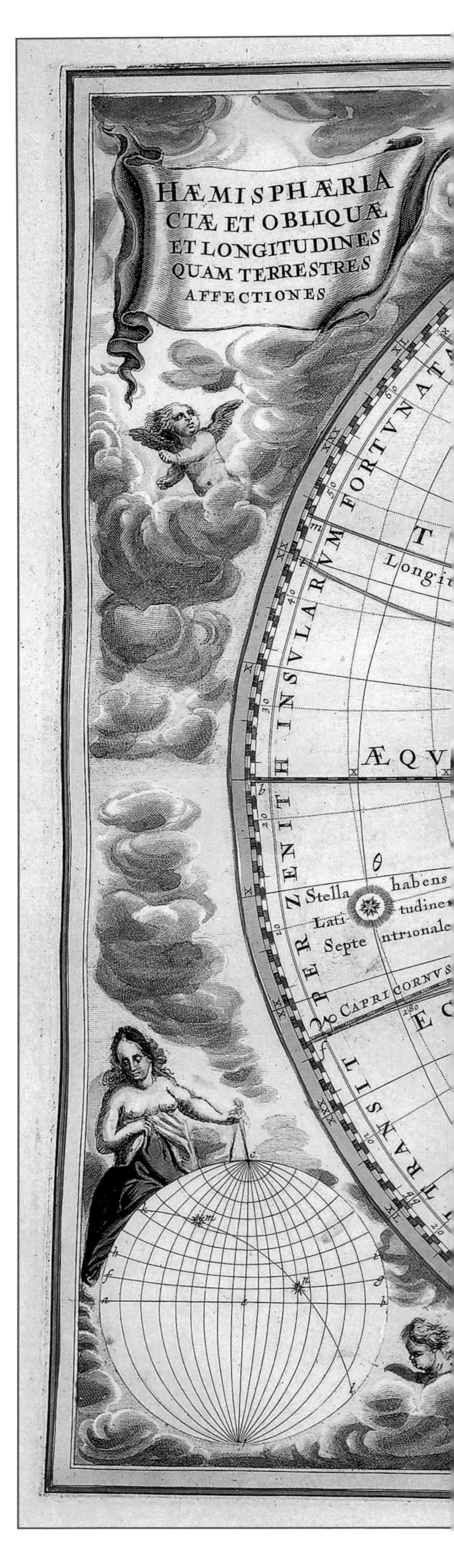

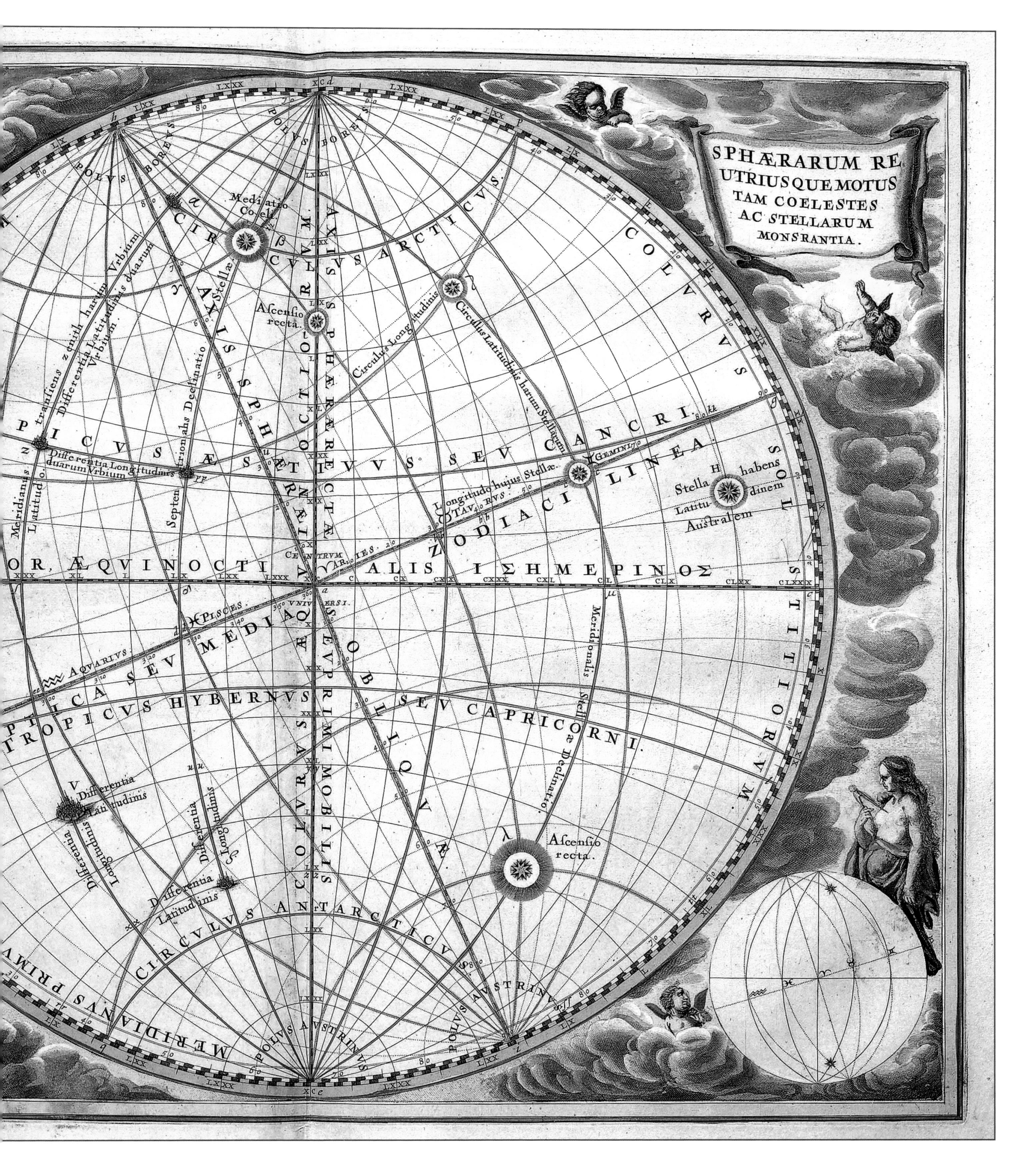

SPHÆRARUM RE
UTRIUSQUE MOTUS
TAM COELESTES
AC STELLARUM
MONSRANTIA.
POLVS BOREVS
CIRCVLVS ARCTICVS.
Mediatio Coeli.
Ascensio recta.
AXIS SPHÆRÆ RECTÆ
AXIS SPHÆRÆ OBLIQVÆ
Circulus Longitudinis
Circulus Latitudinis harum Stellarum
COLVRVS SOLSTITIORVM.
Differentia Latitudinis duarum Vrbium
Meridianus transiens zenith harum Vrbium
Septentrionalis Declinatio
Differentia Longitudinis duarum Vrbium
TROPICVS ÆSTIVVS SEV CANCRI.
Longitudo hujus Stellæ.
TAVRVS
GEMINI
Stella H habens Latitudinem Australem
ZODIACI LINEA
CENTRVM VNIVERSI.
ARIES
ÆQVINOCTIALIS ΙΣΗΜΕΡΙΝΟΣ
PISCES
AQVARIVS
ECLIPTICA SEV MEDIA
TROPICVS HYBERNVS SEV CAPRICORNI.
Meridionalis Stellæ Declinatio.
Differentia Latitudinis
Differentia Longitudinis
Ascensio recta.
CIRCVLVS ANTARCTICVS
POLVS AVSTRINVS
MERIDIANVS PRIMVS

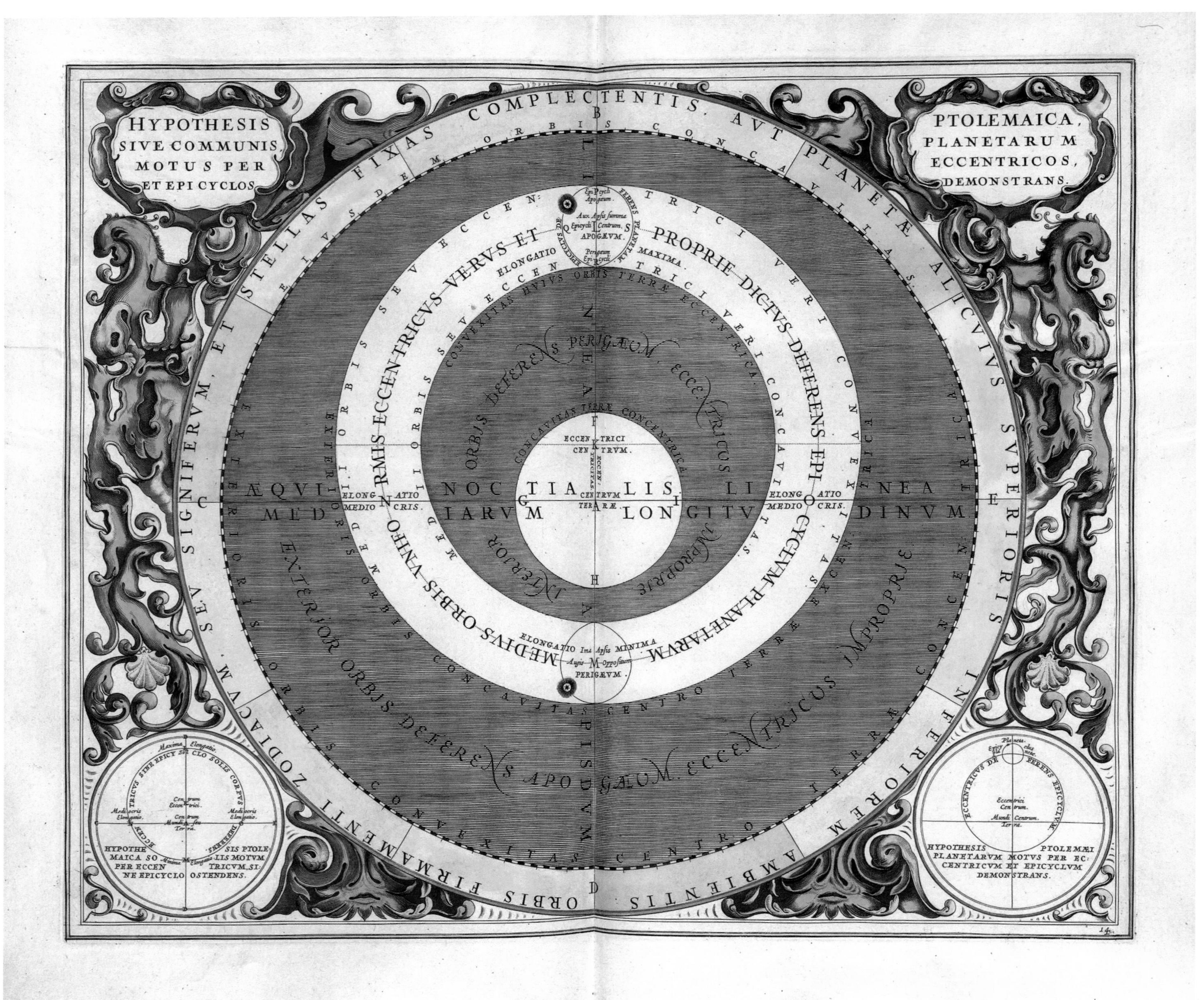

Hypothesis Ptolemaica

The fourteenth plate – titled Hypothesis Ptolemaica, Sive Communis, Planetarum Motus per Eccentricos, et Epicyclos Demonstrans *– represents the hypotheses formulated by Ptolemy regarding the movement of the planets, with their eccentric orbits and epicycles. In the geocentric system, the epicycle is the circle in which the planets and the Sun travel at a constant speed. The center of the epicycle in turn moves in another orbit, called deferent or eccentric, inside which is Earth, in an eccentric position. The theory of the epicycles was formulated in ancient times to explain the apparent movement of the planets, and it was also adopted by Copernicus until another astronomer, Johannes Kepler, discovered that the planetary orbits were elliptic.*

Typus Aspectuum, Oppositionum et Coniunctionum etz in Planetis

This lovely plate, number 15, illustrates the planetary configurations – opposition and conjunction – that may occur in the sky when it is observed from the center of the Earth. Conjunction occurs when two stars or planets have the same longitude or the same right ascension. On the other hand, opposition occurs when one of two celestial bodies lies in the opposite direction of the other with respect to the observer. Our planet is placed in the middle, viewed from above (thus highlighting the Northern Hemisphere) and is surrounded by the Zodiac wheel. A close look at the representation of the continents reveals that in this case as well California looks like an island detached from the rest of North America.

Theoria Solis per Eccentricum sine Epicyclo

The sixteenth plate of Harmonia Macrocosmica *represents the orbit (epicycle) of the Sun around Earth as formulated in ancient times by Ptolemy. In particular, the figure shows the difference between the interval that separates the spring and autumn equinoxes (187 days all told) from the inverse period (178 days). According to Ptolemy, this lack of equilibrium is to be explained by the fact that the Sun's orbit around Earth is not perfectly circular but slightly eccentric; therefore the passage under the equinoctial line (here called* Aequinoctialis seu Colurus Aequinoctiorium*) is inferior and shorter time-wise compared to the Sun's course in the section of the orbit above the same equinoctial line.*

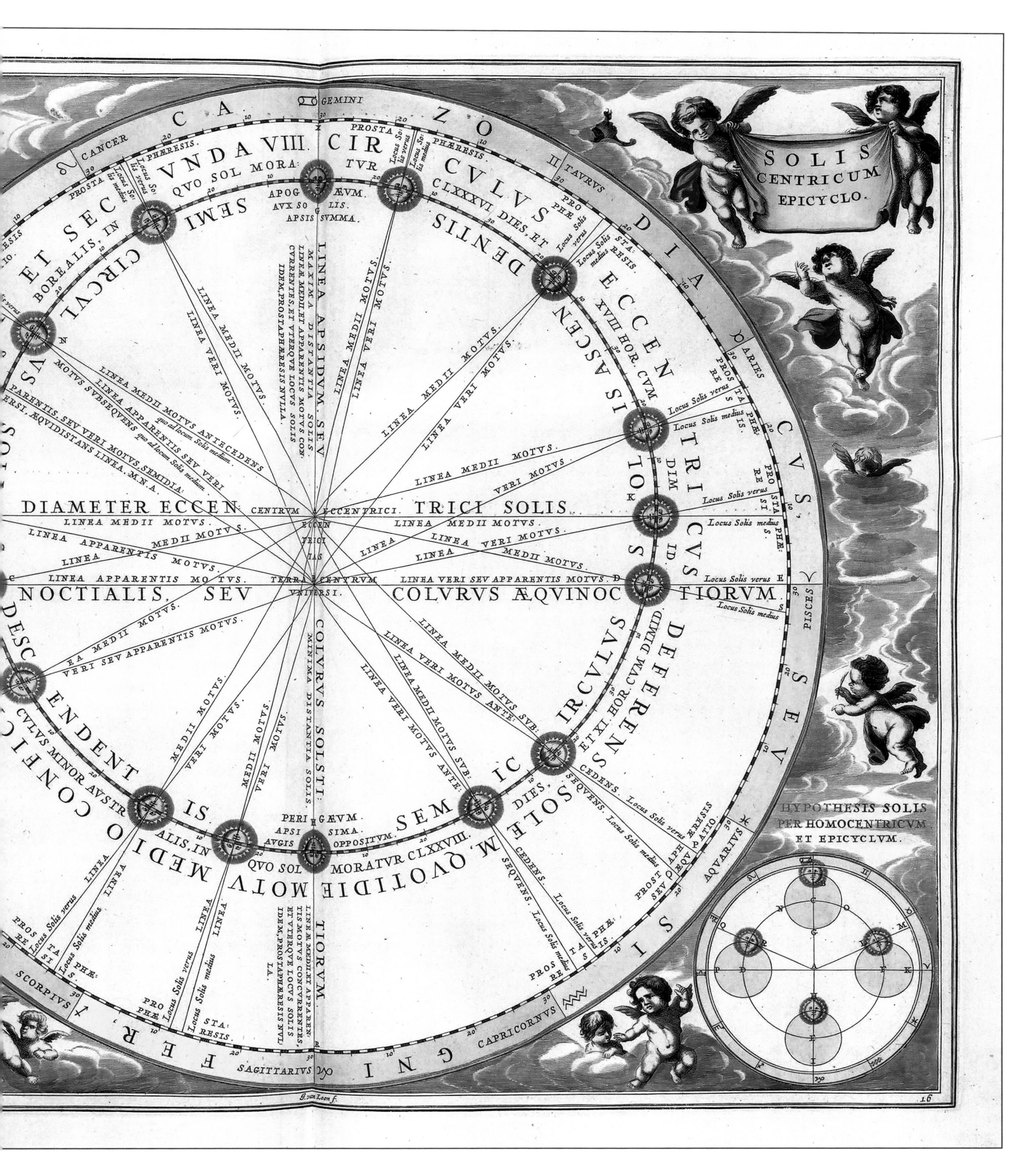

SOLIS
CENTRICUM
EPICYCLO.
HYPOTHESIS SOLIS
PER HOMOCENTRICUM
ET EPICYCLUM.
CANCER
GEMINI
TAURUS
ARIES
PISCES
AQUARIUS
CAPRICORNUS
SAGITTARIUS
SCORPIUS
DIAMETER ECCEN
TRICI SOLIS
CENTRUM ECCENTRICI.
TERRÆ CENTRUM UNIVERSI.
NOCTIALIS, SEU
COLURUS ÆQUINOC TIORUM.
LINEA APSIDUM, SEU MAXIMA DISTANTIA SOLIS.
COLURUS SOLSTI: MINIMA DISTANTIA SOLIS.
APOG ÆUM. AUX SOLIS. APSIS SUMMA.
PERI GÆUM. APSI SIMA. AUGIS OPPOSITUM.
QUO SOL MORA: TUR
CLXXXVI DIES, ET XVIII HOR. CUM DIMID.
QUO SOL MORATUR CLXXVIII. DIES, ET XI. HOR. CUM DIMID.
LINEA MEDII MOTUS.
LINEA VERI MOTUS.
LINEA APPARENTIS MOTUS.
LINEA VERI SEU APPARENTIS MOTUS.
Locus Solis verus
Locus Solis medius
G. van Loon f.
16

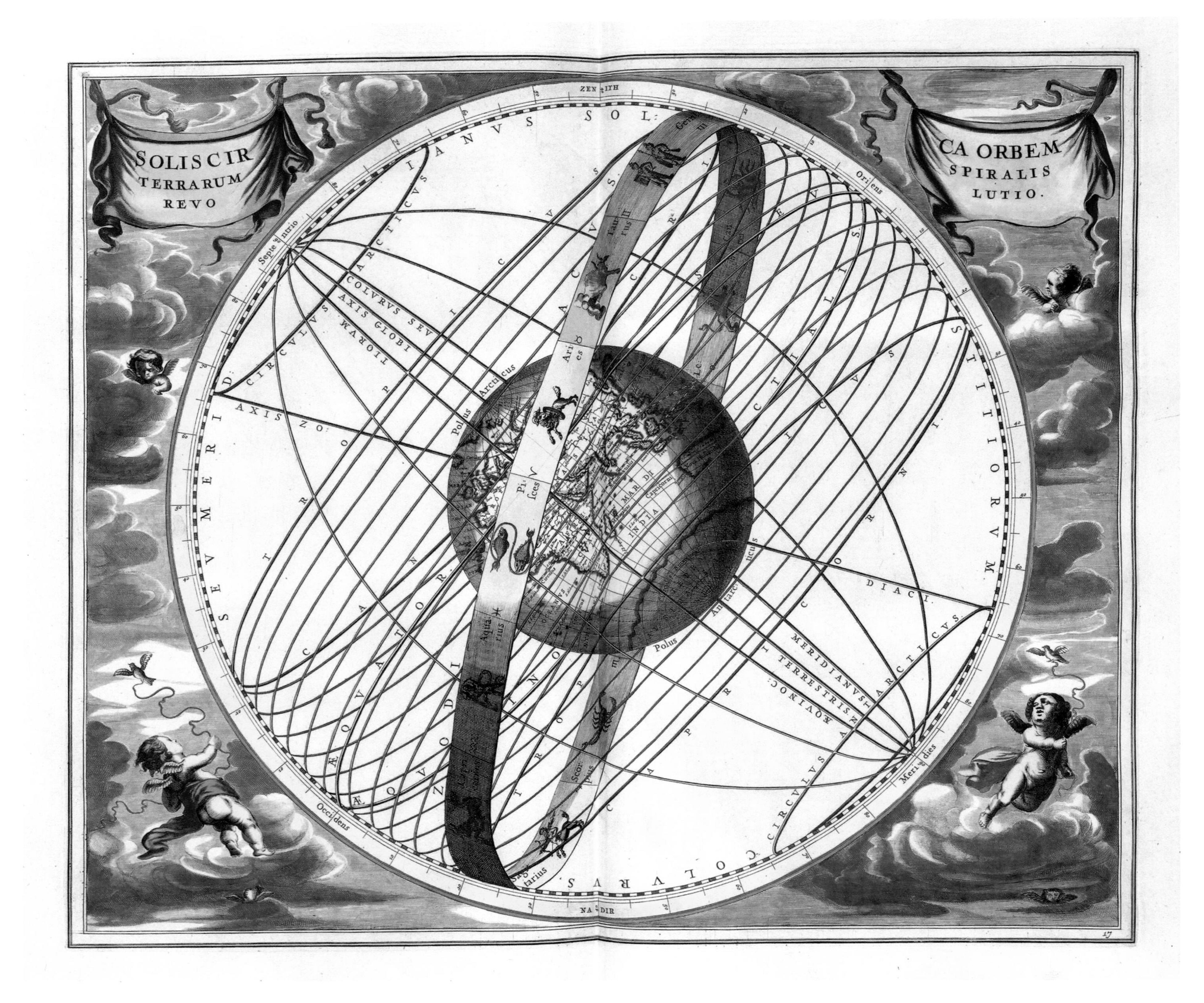

Solis Circa Orbem Terrarum Spiralis Revolutio

Here Cellarius represents the apparent spiral revolution of the Sun around Earth. The movement of the former is described thanks to the contribution of a circular cornice that registers its ever-changing positions, from the Zenith (the point in the celestial sphere perpendicular to the horizon and directly above the observer) to the Nadir (the point directly below the observer). This plate is number 17 of the Atlas.

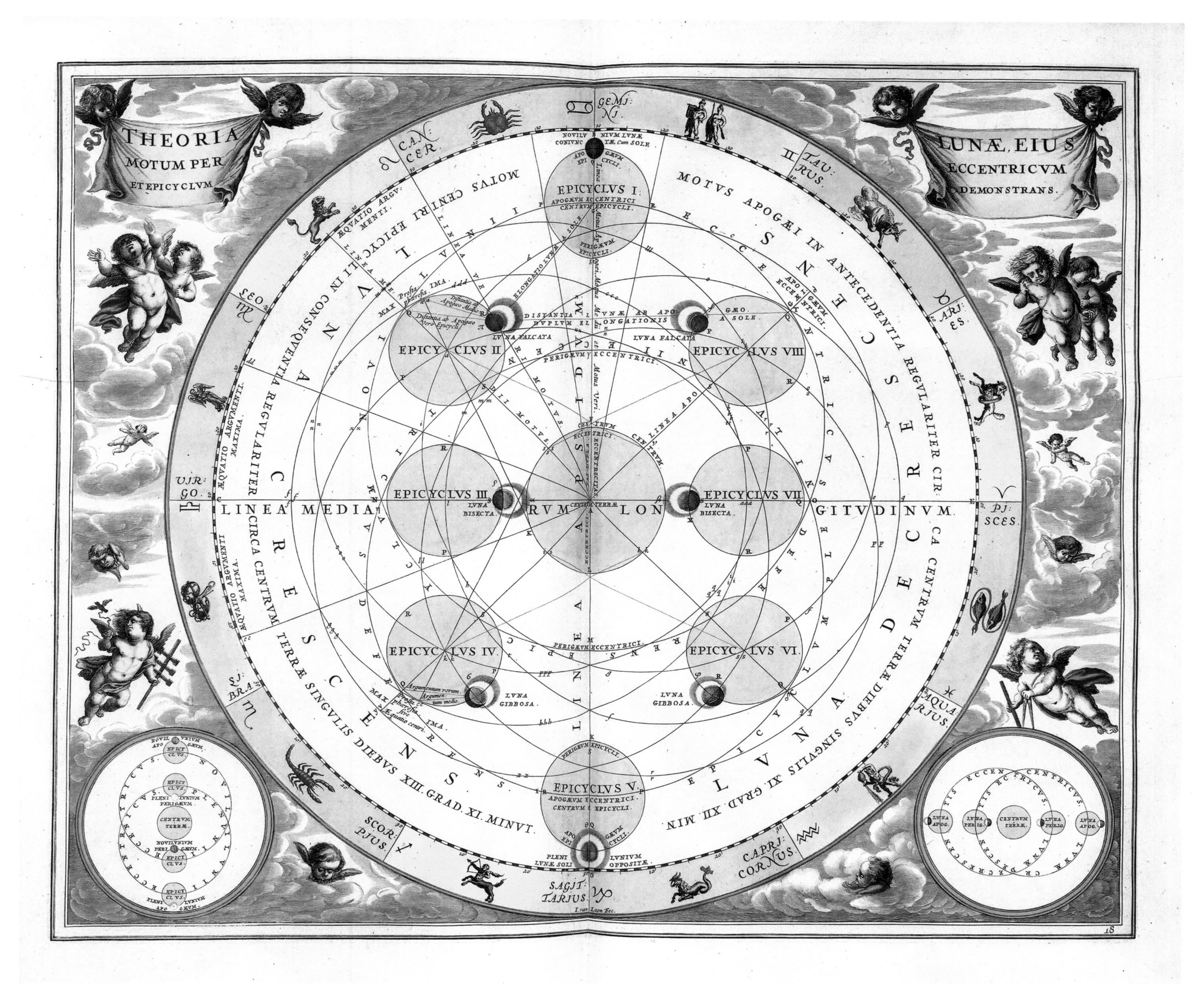

Theoria Lunae, Eius Motum per Eccentricum et Epicyclum Demonstrans

Plate number 18 illustrates the movement of the Moon around Earth: the epicycles are depicted with a series of eight circles arranged around the axis of orbital revolution. Thus one can also appreciate the lunar phases, from the new moon (above) to the full moon (below), passing through the intermediate waning and waxing phases – crescent, bisected, and gibbous. The lower corners of the plate have diagrams illustrating the eccentric nature of the lunar orbit around Earth.

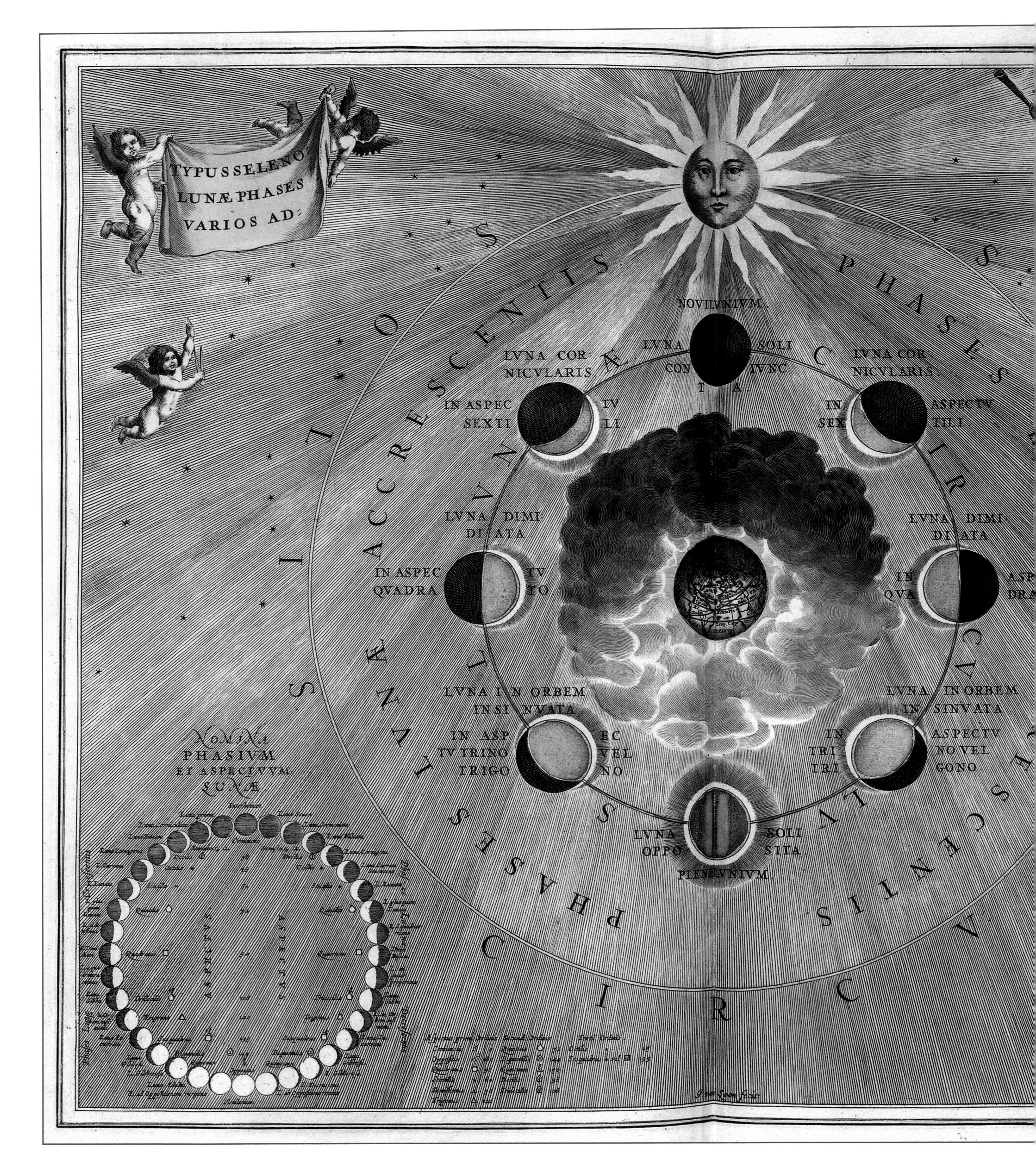

TYPUSSELENO
LUNÆ PHASES
VARIOS AD:
NOVILUNIUM.
LUNA SOLI CONIUNCTA.
LUNA CORNICULARIS
IN ASPECTU SEXTILI
LUNA DIMIDIATA
IN ASPECTU QUADRATO
LUNA IN ORBEM INSINUATA
IN ASPECTU TRINO VEL TRIGONO
LUNA SOLI OPPOSITA.
PLENILUNIUM.
ACCRESCENTIS
PHASES
NOMINA
PHASIUM
ET ASPECTUUM
LUNÆ

Typus Selenographicus Lunae Phases et Aspectus Varios Adumbrans

In this spectacular plate (number 19), Cellarius once again represents the movement of the Moon, this time placing emphasis on its phases and on how it appears to those observing it. The work shows how the portion of the Moon's surface changes and is more or less visible, depending on its position with respect to the Sun, whose rays illuminate it. The illustration below left, titled Nomina Phasium et Aspectuum Luna, *is a detailed representation of the 36 different lunar sub-phases with their respective names, while the diagram below right (*Phases Lunae Respectu Solis et Oculi*) describes the twelve principal phases as they appear when viewed from the Sun, or vice versa when seen by an observer on Earth (*respectu Solis et oculi*).*

Theoria Trium Superiorum Planetarum

Plate 20 is an efficacious illustration of the movement of the three superior planets: Mars, Jupiter and Saturn. The graphs illustrate the theory of the epicycles and deferents as formulated by Ptolemy. Already in plate fourteen Cellarius had proposed an illustration of the theory of epicycles in general, while in this illustration he goes into detail, showing how, in Ptolemy's view, the planets revolve around Earth in imperfectly circular orbits. The plate is admirable for the equilibrium and symmetry of the composition and the quality of execution, which is enhanced by the usual cherubs that stand out against the leaden sky.

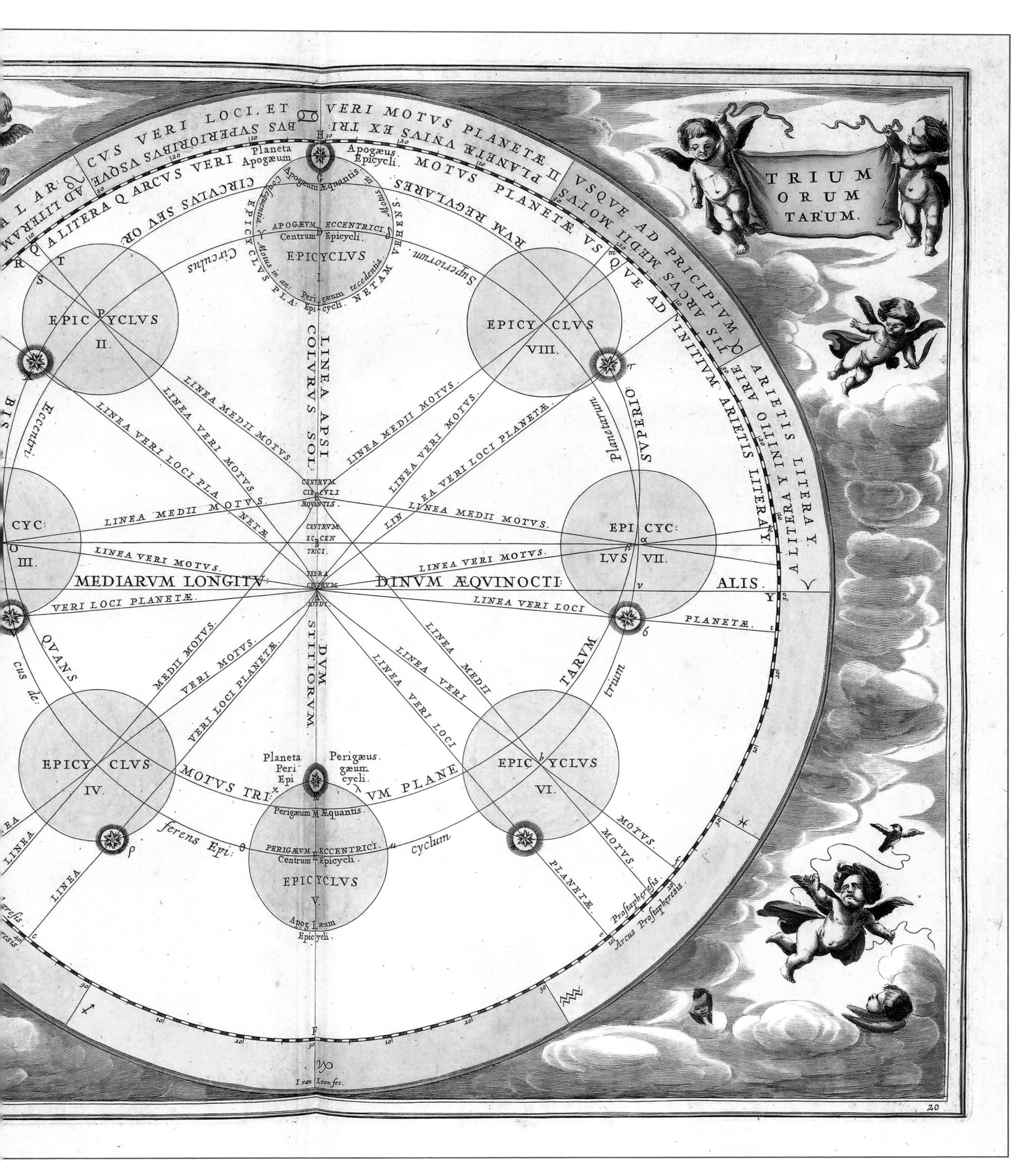
TRIUM
ORUM
TARUM.
CVS VERI LOCI, ET
VERI MOTVS PLANETÆ II
VSQVE AD PRICIPIVM
ARIETIS LITERA Y.
ALITERA Y INITIO ARIE
Planeta Apogæum
Apogæus Epicycli.
Apogæum Æquantis.
APOGÆVM ECCENTRICI.
Centrum Epicycli.
EPICYCLVS I.
Perigæum Epicycli.
EPICYCLVS II.
EPICYCLVS VIII.
LINEA APSI COLVRVS SOL
LINEA MEDII MOTVS.
LINEA VERI MOTVS.
LINEA VERI LOCI PLANETÆ.
CENTRVM CIRCVLI ÆQVANTIS.
CENTRVM ECCENTRICI.
TERRA CENTRVM MVNDI.
MEDIARVM LONGITV
DINVM ÆQVINOCTI
ALIS.
EPI CYC LVS VII.
LINEA VERI LOCI PLANETÆ.
DVM STITIORVM
EPICYCLVS IV.
EPICYCLVS VI.
Planeta Perigæus.
Perigæum Epicycli.
Perigæum Æquantis.
PERIGÆVM ECCENTRICI.
Centrum Epicycli.
EPICYCLVS V.
Apogæum Epicycli.
MOTVS TRI VM PLANE
ferens Epi cyclum
Prosthaphæresis.
Arcus Prosthaphæresis.
I. van Loon fec.
20

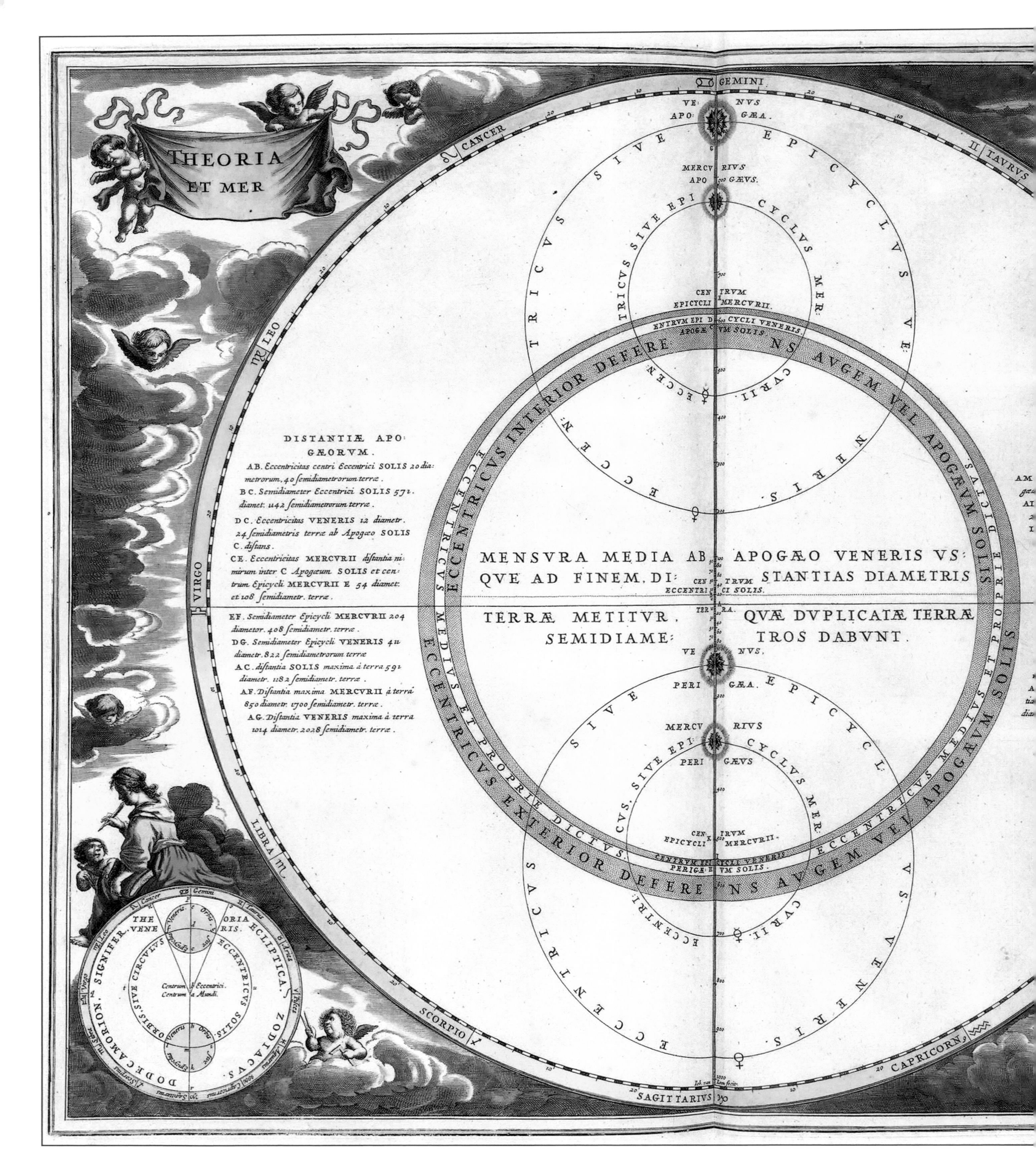
THEORIA
ET MER
GEMINI
CANCER
TAVRVS
LEO
VIRGO
LIBRA
SCORPIO
SAGITTARIVS
CAPRICORN
VENVS APOGÆA.
MERCVRIVS APOGÆVS.
CENTRVM EPICYCLI MERCVRII.
MENSVRA MEDIA AB APOGÆO VENERIS VSQVE AD FINEM, DISTANTIAS DIAMETRIS
TERRÆ METITVR, QVÆ DVPLICATÆ TERRÆ SEMIDIAMETROS DABVNT.
VENVS.
PERIGÆA.
MERCVRIVS
PERIGÆVS
CENTRVM EPICYCLI MERCVRII.
ECCENTRICVS INTERIOR DEFERENS AVGEM VEL APOGÆVM SOLIS
ECCENTRICVS EXTERIOR DEFERENS AVGEM VEL APOGÆVM SOLIS
ECCENTRICVS MEDIVS ET PROPRIE DICTVS
DISTANTIÆ APOGÆORVM.
AB. Eccentricitas centri Eccentrici SOLIS 20 diametrorum, 40 semidiametrorum terræ.
BC. Semidiameter Eccentrici SOLIS 571. diamet: 1142 semidiametrorum terræ.
DC. Eccentricitas VENERIS 12 diametr. 24 semidiametris terræ ab Apogæo SOLIS C. distans.
CE. Eccentricitas MERCVRII distantia nimirum inter C Apogæum SOLIS et centrum Epicycli MERCVRII E 54 diamet: et 108 semidiametr. terræ.
EF. Semidiameter Epicycli MERCVRII 204 diametor. 408 semidiametr. terræ.
DG. Semidiameter Epicycli VENERIS 411 diametr. 822 semidiametrorum terræ
AC. distantia SOLIS maxima à terra 591 diametr. 1182 semidiametr. terræ.
AF. Distantia maxima MERCVRII à terra 850 diametr. 1700 semidiametr. terræ.
AG. Distantia VENERIS maxima à terra 1014 diametr. 2028 semidiametr. terræ.
THEORIA VENERIS.
ECLIPTICA.
ZODIACVS.
SIGNIFER.
DODECAMORION.
ECCENTRICVS SOLIS
ORBIS, SIVE CIRCVLVS
Centrum Eccentrici.
Centrum Mundi.

Theoria Veneris et Mercurii

Like the preceding plate, the twenty-first continues the illustration of the movement of the planets based on the theory of epicycles. However, in this case the protagonists are Venus and Mercury. On either side of the diagrams, Cellarius inserts the calculations of the apogees (above) and perigees (below) of Venus and Mercury, that is, the maximum and minimum distances of the orbit of the planets from the central point of reference. Here again the plate is notable for its high-quality execution and aesthetic and formal equilibrium; and the finishing touch is the typical baroque decoration featuring female amanuenses, astronomical instruments and cherubs.

Coeli Stellati Christiani Haemisphaerium Prius

True to his project of providing a complete, systematic and unbiased picture of the astronomical theories considered valid in his time, Cellarius dedicated two plates in his atlas to representations of the sky proposed by Julius Schiller, that is, a Christian interpretation of the constellations. This illustration (plate number 22) represents the first of the two hemispheres, with Schiller's 'new' constellations – which were inspired by the Old and New Testament and the history of Christianity – replacing the traditional names based on classical mythology. Among the many biblical figures and icons are St. Michael Archangel, the Magi, St. Sylvester, St. Benedict, Noah's Ark, King David, the Paschal Lamb, Abraham and Isaac, and the Ark of the Covenant.

STELLATI
ANI HÆ
RIUM PRIUS.
Definitor seu Colurus Æquinoct. et Axis Mundi.
Circulus Polaris Sept:
Definitor
seu
Colurus
Tropicus Æstivus Solstiti:
Æquinoctialis. vel
Ecliptica.
S.Michael
Navicula S.Petri
S.Ioannes. Al.Cancer.
S.Siluester
TRES REGES.
Flagellum XPI. Al.Coma Berenices.
Corona XPI.
S.Benedictus cum Spina Eiusd.
S.Thomas Al.Leo.
S.Iacobus minor
Area Fœderis Al.Crater et Cornus
Jordanis.
Area Noë. Al.Argo Navis
S.Abraham et Isaac. Al.Centaurus.
S.Israel seu Iacob. Al.Lupus.
S.Bartholomæus Al.Scorpio.
Signum Tau.
Polus Mundi Merid.
Circulus Polaris Merid.
Axis Mundi et
Definit: aut Colur: Æquinoct.
22

COELI
CHRISTI
SPHÆRIUM
Definitor seu Colurus Æquinoct. et Axis Mundi.
Circulus Polaris Sept.
Axis Eclipticæ Polus Eclipticæ sive Zodiaci Boreus.
Polus Eclipticæ Boreus.
Tropicus Æstivus.
Æquinoctialis.
Ecliptica.
S.Stephanus Al. Cepheus
S.Michael Al. Ursa maior.
Nauicula S.Petri.
S.S.Innocentes Alias Draco.
Praesepe XPI. Alias Lyra.
S.Crux XPI etiam S.Helena Al. Cygnus
S.Gabriel Al. Pegasus
S.Andreas. Al. Taurus.
S.Catharina Al. Aquila cum Antinoo.
Hydria Chananæa Al. Delphin
Rosa Mystica. Al. Equiculus.
S.Matthias Al. Pisces
S.Petrus Al. Aries.
S.Iudas Thaddæus. Alias Aquarius.
S.Ioachimus et Anna Al. Cetus
S.Simon. Al. Capricornus.
S.Matthæus Al. Sagitarius
Hydria Sareptana Al. Piscis Notius.
S.Aaron Al. Grus
S.Iob. Al. Indus et Pavo.
Circulus Polaris Merid.
Polus Mundi Merid.
Axis Eclipticæ Polus Eclipticæ sive Zodiaci Austrinus.
Axis Mundi et Definit. aut Colur. Æquinoct.

Coeli Stellati Christiani Haemisphaerium Posterius

Plate number 23 is paired with the preceding one in that it represents the second of the hemispheres with the 'Christian' constellations conceived by Schiller as replacements of the pagan ones. Among these are St. Stephen, St. Helena, St. Catherine, Job, St. Jerome and Mary Magdalene, as well as the Archangels Gabriel and Raphael. In the Zodiac the classical constellations are given the names of the Twelve Apostles. However, it must be said that Cellarius remains loyal to his all-inclusive and didactic goal by maintaining the original pagan names of the constellations, which he places under the Christian ones. From an aesthetic standpoint the two plates are a true 'triumph' of baroque taste: the lack of 'empty spaces' is almost asphyxiating, in line with the horror vacui *typical of the art in this period.*

Haemisphaerium Stellatum Boreale Antiquum

*Plate number 24, one of the most famous in Cellarius's Atlas, reproduces the stars of the Northern Hemisphere as conceived in the past. The constellations are the classic ones, which as always drew inspiration from ancient mythology. But the stars, divided into classes of six different magnitudes or degrees of brightness, are depicted as they must have appeared in 1660. Besides the Ptolemaic constellations, the plate also has those dedicated to the Tigris, Euphrates and Jordan rivers, and to Antinoüs, Bee (*Apes*), Giraffe (*Gyraffa*), Lesser Crab (*Cancer Minor*) and Sagitta or Southern Arrow (*Sagitta Australis*) – which were introduced by Petrus Plancius in 1613.*

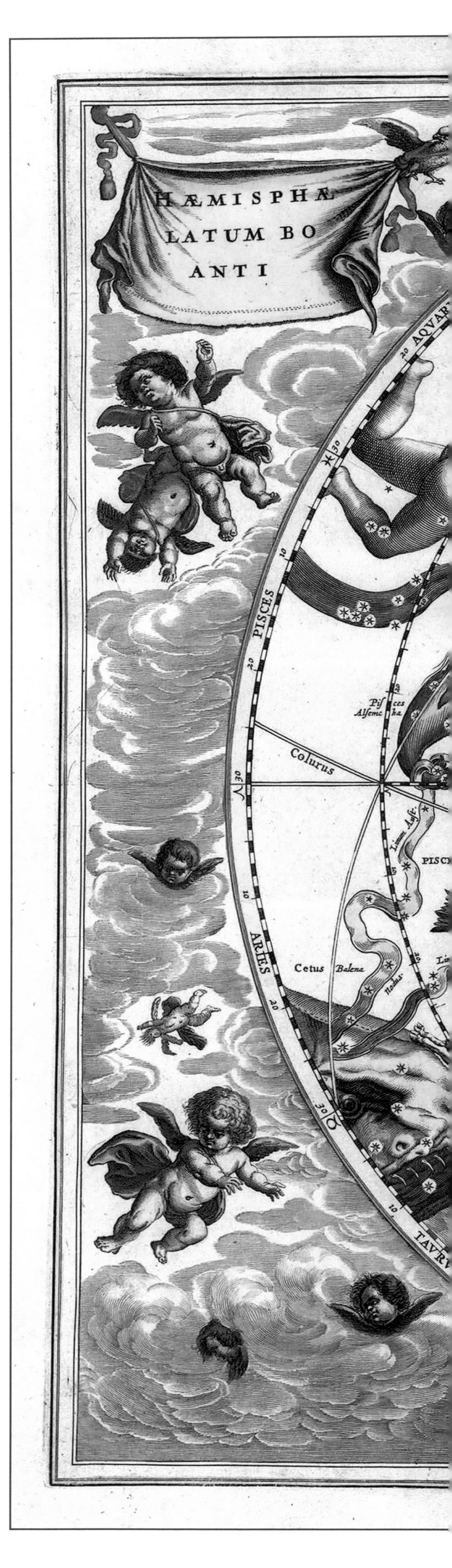

RIUM STEL
REALE
QVUM.
CAPRICORNVS
SAGITTARIVS
SCORPIVS
LIBRA
VIRGO
LEO
CANCER
GEMINI
SCORPIO
Capricornus
Zodi
acus
Sagittarius
Serpentarius
Antinous
Aqvarius
Aquila
Equuleus
Delphinus
Lyra
Engonasi
Serpens
Circul
Solstitiorum
Æquinoctialis
Æquinoctiorum
Libra
Corvus
Virgo
Bootes
Arcturus
Corona Sept.
Cygnus
Pegasus
Draco
Coma Berenices
Andromeda
Cepheus
Vrsa minor
Vrsa major
Cassiopeia
Aries
Triangulum
Apes
Camelopard
Arcticus
Auriga
Leo, Alased
Cancri
Perseus
Tropicus
Taurus
Gemini
Cancer
Colurus
Orion
Aldebaran

HEMISPHÆ
ALIS COELI
SPHÆRI
GRA-
Cassiopeia
Cepheus
Draco
Cygnus
Ursa Major
Ursa Minor
Engonasi
Corona
Serpens
Ophiuchi
Bootes
Bubulcus
VIRGO
Æquinoctialis
Coma Berenices
LEO
Cor Leonis
CANCER
Crater
Nova Hollandia
Corvus
Hydra
Centaurus
Argo
Navis

Hemisphaerii Borealis Coeli et Terrae Sphaerica Scenographia

In this map, the twenty-fifth plate, Cellarius offers a 'scenographic' picture of the Northern Hemisphere as a three-dimensional sphere. If observed carefully, one can see below the constellations the corresponding portion of our planet with its continents, oceans, seas and major rivers. This is a prodigious work of synthesis whose objective is to portray Earth and the sky in the most thorough manner possible. In fact, Cellarius achieves this aim to the utmost if we stop to consider four of the following five plates – that is, numbers 26, 28 and 29, besides this one – as parts of a whole that represent, albeit in a different manner, the starry sky and Earth as a single entity.

Haemisphaerium Stellatum Australe Antiquum

*This plate, number 27 of the series, is to be coupled with number 24 since it represents the Southern Hemisphere as viewed by the ancients. The 'classical' constellations that Ptolemy was already acquainted with are here integrated by 'new' ones that had been recently introduced by Petrus Plancius: Chamaeleon, Goldfish, Phoenix, Crane, Small Water Snake, Indian, Fly (Bee), Peacock, Flying Fish, Southern Triangle, Toucan, and Bird of Paradise. All these constellations were conceived after the exploration of unknown or little known lands carried out by navigators in the Southern Hemisphere had provided further data. These constellations, together with Dove (*Columba*, introduced by Plancius in 1592), had already been depicted in 1603 by Johann Bayer in his* Uranometria.

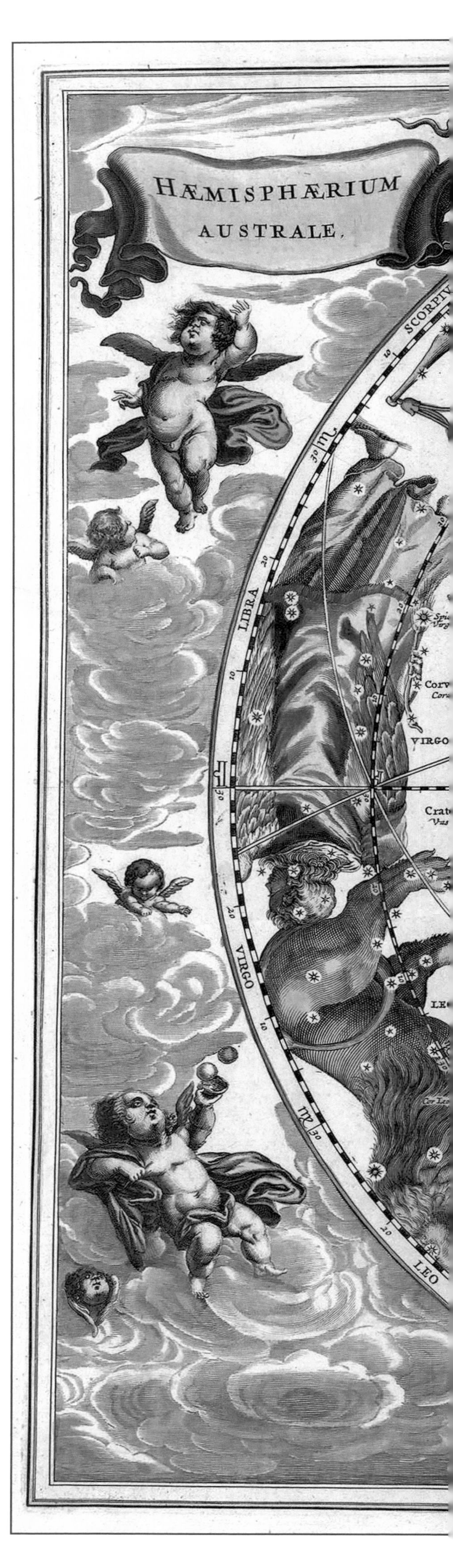

STELLATUM ANTIQUUM.
SAGITTARIUS
CAPRICORNUS
AQUARIUS
PISCES
ARIES
TAURUS
GEMINI
CANCER
Serpentarius Ophiuchus
SCORPIO
Sagitta Austr.
Corona Aust.
Fera Lupus
Ara
Pavo
Indus
Grus
Pisces Aust.
Toucan
Phœnix
Capri cornus
Urna
Centaurus Chiron
Cruzero
Musca
Apis Indica
Triangulum aust.
Colurus
Chameleon
Arca Noachi
Æquinoctialis
Dorado
Cetus Balena
Fluvius Eridanus
Canobus Sohel
Columba Noachi
Tropicus
Hydra
Circulus
Lepus Alar: nebet
Canis major Sirius
Monoceros Unicornis
Canis minor Porcoyn
CANCER Asartan
Asinus Bore:
Cancer minor
Pollux alijs Hercules
GEMINI Algense
Orion
TAURUS Altor
PISCES Alsemcha
Zodi:
27

Haemisphaerium Stellatum Boreale cum Subiecto Haemisphaerio Terrestri

In plate 26 Cellarius depicts the Northern Hemisphere supported by two mythological figures: at right, the hero and demigod Hercules, and at left, Atlas. Unlike plate 25, with the same subject, the representation of the starry northern sky and the continents underneath it is turned so that the constellations and lands that were mostly hidden in the preceding plate are visible in this one.

Haemisphaerium Scenographicum Australe Coeli Stellati et Terrae

The penultimate plate (number 28) is a spectacular rendering of the Southern Hemisphere, both terrestrial and celestial, and continues the complete mapping of Earth and the sky already begun in plates 25 and 26. Here again the representation is effected by utilizing the only possible viewpoint, with the hypothetical observer outside the illustration.

Haemisphaerium Stellatum Australe Aequali Sphaerarum Proportione

The twenty-ninth and last plate of Cellarius's striking atlas represents the Southern Hemisphere as a celestial sphere held up by Hercules and Atlas. This map completes the preceding one by highlighting the constellations, which were not visible in plate number 28. Once again, the elaborate depiction fully complies with the standards and taste of baroque art. However, compared to its 'twin' plate, which represents the Northern Hemisphere, here the land is less extensive and more concentrated, so the plate is not overflowing with figures and is thus much 'lighter', airy and enjoyable.

RIUM STEL
STRALE ÆQUALI
PROPORTIONE.
Lucida Lyræ
Tropicus
Engonasi
Ras Alguethi
Canceri
Delphinus
Aquila
Equuleus
Serpens Ophiuchi
Antinous
Serpentarius
Circulus
LIBRA
SAGITTARIUS
AQUARIUS
SCORPIO
Fera Lupus
Ara
Pavo
Grus
Centaurus
Toucan
Phœnix
Hydra

Johannes Hevelius

(1611-1687)

Johannes Hevelius (or, in his mother tongue, Polish, Jan Heweliusz) is a decisive figure in the history of astronomy because he founded lunar topography (the science that studies and charts the surface of the Moon) and also because he described eleven new constellations after the 48 already listed by Ptolemy in his Almagest *and those introduced by Bayer. Moreover, of these, seven are still recognized. His most famous work is* Prodromus Astronomiae, *which was published posthumously at Danzig in 1690 by his second wife Elisabetha Koopman, who collaborated with him in his observations of the sky and is therefore considered the first woman astronomer in history.*

Hevelius was born on 28 January 1611 in Danzig to a family of wealthy brewers of Czechoslovakian origin. After studying jurisprudence at Leiden and traveling in England and France, in 1634 Hevelius returned to Danzig, where he took an active part in local politics, becoming a city councilor, and then mayor in 1651. He was quite adept at managing the family's brewery business, but soon found that his true passion was astronomy. In 1641 he built an observatory on the roof of his house, equipping it with state-of-the-art instruments, including a 'tubeless' telescope (built with a series of refracting lenses mounted on a long wooden structure) whose focal length was 150 feet (45 meters). The results of his daily observations were published in 1647 in Selenographia, sive Lunae Descriptio, *accompanied by extremely precise cartography consisting of four general maps and 40 excellent engravings, each of which contains a different aspect of the lunar phases. This was followed in 1656 by* Dissertatio de Nativa Saturni facie, *a treatise in which Hevelius described Saturn. The year 1661 saw the publication of* Mercurius in Sole visus, *in which the astronomer described the transit of Mercury in front of the Sun, which he himself observed on May 3; he also made a very precise measurement of its diameter (his estimate differs less than 0.5 seconds from its real length). From the height of his well equipped observatory – which was visited by the King of Poland, John II Casimir, by his consort Maria Gonzaga, as well as by Edmond Halley, who went there on behalf of the Royal Society of London (of which Hevelius was a member since 1664) – the Polish astronomer devoted a great deal of energy to the observation of the stars and comets. In* Historiola Mirae *(1662) he described the variable star Omicron Ceti, which he christened Mira ('wonderful' in Latin) since it was so different from the other known stars. In* Prodromus Cometicus *(1665) he examined the large comet that appeared on 14 December 1664, depicting it in 28 drawings that followed its evolution until it disappeared from sight on 18 February 1665. The results of his research on comets were crowned in 1668 by his* Cometographia, *which contained descriptions and an analysis of four comets he had discovered, whose movement around the Sun he theorized as having a parabolic trajectory.*

Tragedy occurred on 26 September 1679: Hevelius's observatory was totally destroyed by fire, together with his precious instruments, printing press and library. He himself described the terrible event in the preface to his Annus climactericus *(1685). The flames also consumed most of the copies of the second volume of* Machina coelestis, *a long treatise that described his instruments and the data he had gathered during his observations; the books had just been printed and were piled up in his print shop, which was also destroyed by the fire. With great difficulty Hevelius managed to get his study back in shape, but the trauma overwhelmed him and from that day on he was never the same. He died on 28 January 1687 in his home town of Danzig, after a long illness. He was buried in St. Catherine's Church, where his remains still lie.*

A portrait of Johannes Hevelius by Daniel Schultz (oil on canvas, 1677; Danzig Library of the Polish Academy of Sciences).

Prodromus Astronomiae
(1690)

This work is divided into three volumes. This first contains the preface (*Prodromus*) and a series of observations concerning the author's methodology; the second (*Catalogus Stellarum Fixarum*) is a catalog of 1564 stars, listed according to their respective constellations; the third, known as *Firmamentum Sobiescianum, sive Uranographia*, is a star atlas with reproductions of the constellations in 56 splendid plates executed by Hevelius himself (except for three, which were the work of the engravers Andreas Stech and Carolus de la Haye). The volume is dedicated to John III Sobieski, king of the Polish-Lithuanian Commonwealth from 1674 to 1696 and the hero of the 1683 Battle of Vienna, during which he defeated the Ottomans. In acknowledgement of this latter feat, Hevelius gave the following name to one of the eleven constellations that he added in his volume to Ptolemy's 48: Scutum Sobiescianum ('Shield of Sobieski', now known simply as Scutum). The other constellations the astronomer introduced, and that are still recognized, are Canes venatici (Hunting Dogs), Leo Minor (Lesser Lion), Lynx, Sextans (Sextant), Lacerta (Lizard) and Vulpecula (Fox).

Prodromus Astronomiae is an extraordinary work from both an aesthetic and scientific standpoint. The splendid presentation of the exceptionally numerous graphic details make the plates of the constellations nothing more or less than true works of art. And the data are up-to-date, since they include information from Kepler's *Tabulae Rudolphinae* (1627) as appraised and revised by Brahe, as well as the very recent measurements taken in 1676 by Edmond Halley on the island of St. Helena, in the southern Atlantic Ocean, which are integrated by Hevelius with his own personal observations.

Frontispiece

The frontispiece of the third part of the Prodromus Astronomiae, *the* Firmamentum Sobiescianum, sive Uranographiae, *presents the muse Urania in the middle of a line of great astronomers, including Copernicus (the first at right), Ptolemy (the first to the left of Urania) and Tycho Brahe (the penultimate at left). At the bottom of the plate is a cartouche with Hevelius's dedication to the king of the Polish-Lithuanian Commonwealth, John III Sobieski.*

Soli DEO Gloria
Magna Opera
Salus et Honor et Gloria et Virtus Dn. DEO nostro.
Sanctus Sanct Sanctus Dn. DEUS Omnipotens.
Dignus es Domine qui accipias Gloriam et Honorem.
Gratiæ et Honor vobis debentur immortales, quòd tam molimine et industriâ, hos immensos exantlastis Labores.
URANIA
DEO debetur Gloria.
Timochares. Hipparchus
Ptolomæus Albategnius. Princeps Hass. Regiomont. Copernicus.
Quæcunque Divina concessit Benignitas, hæc submisse siste, atque offero, Vestroque sublimi committo judicio.
Catalog. Fixarum
FIRMAMENTUM SOBIESCIANUM
sive
URANOGRAPHIA
Joh. Hevelii.
GEDANI
Anno 1687.
Carolus de la Haye. sculp.

Hemisphaerium Firmamenti Sobiesciani Boreale

This beautiful plate depicts the Northern Hemisphere sky with all the visible constellations: besides the traditional ones already catalogued by preceding astronomers, there are the new constellations introduced by Hevelius himself: Scutum Sobiescianum (Shield), Canes venatici (Hunting Dogs), Leo Minor (Lesser Lion), Lynx, Lacerta (Lizard) and Vulpecula (Fox).

Hemisphaerium Australe

This plate illustrates the visible constellations in the Southern Hemisphere. Hevelius added Sextans (Sextant), which he introduced, to those already known. In certain editions of Firmamentum *both these plates representing the sky were colored by hand, which makes us appreciate their beauty even more and become aware of the variety of the details.*

Longitudo.
Draco.
URSA MAJOR
Lynx.
Fig. D.
Latitudo.
Latitudo.
Longitudo.

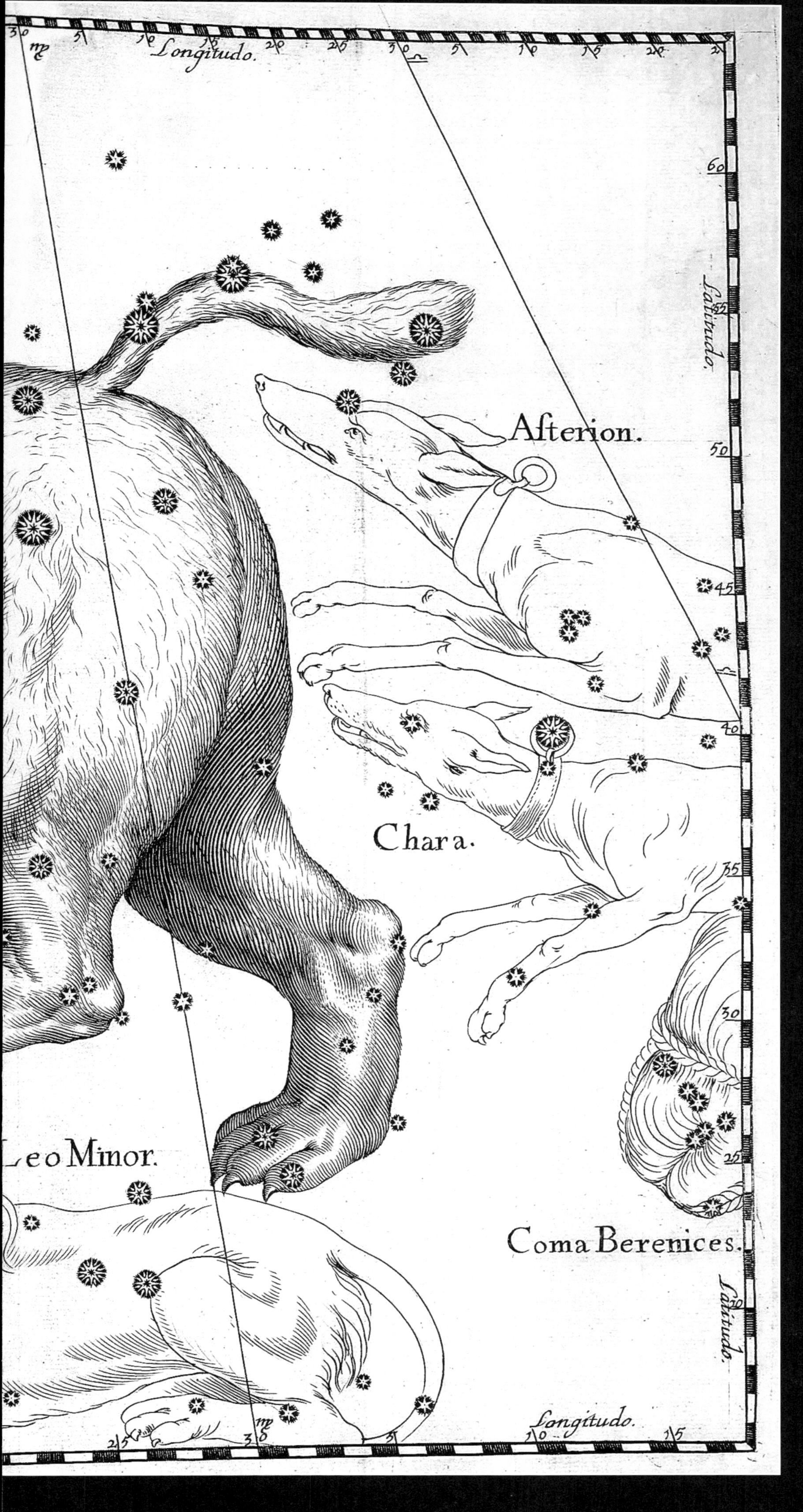

Ursa Maior

In his Atlante Celeste *(Celestial Atlas), the scholar Felice Stoppa points out how Hevelius decided to represent the constellations "viewed from outside the ideal sphere where they lie, a choice that led to result of overturning the position and thus removing the correspondence with the real sky." This is the reason why the Dutch astronomer was contested by later authors, first of all John Flamsteed, who considered this choice misleading. The Ursa Major is no exception, and in the drawing it also has a rather threatening aspect. Below it one can see the Leo Minor constellation, which was introduced by Hevelius himself and is still officially recognized: this is a lion cub, in a squatting position, that accompanies the zodiacal Leo constellation.*

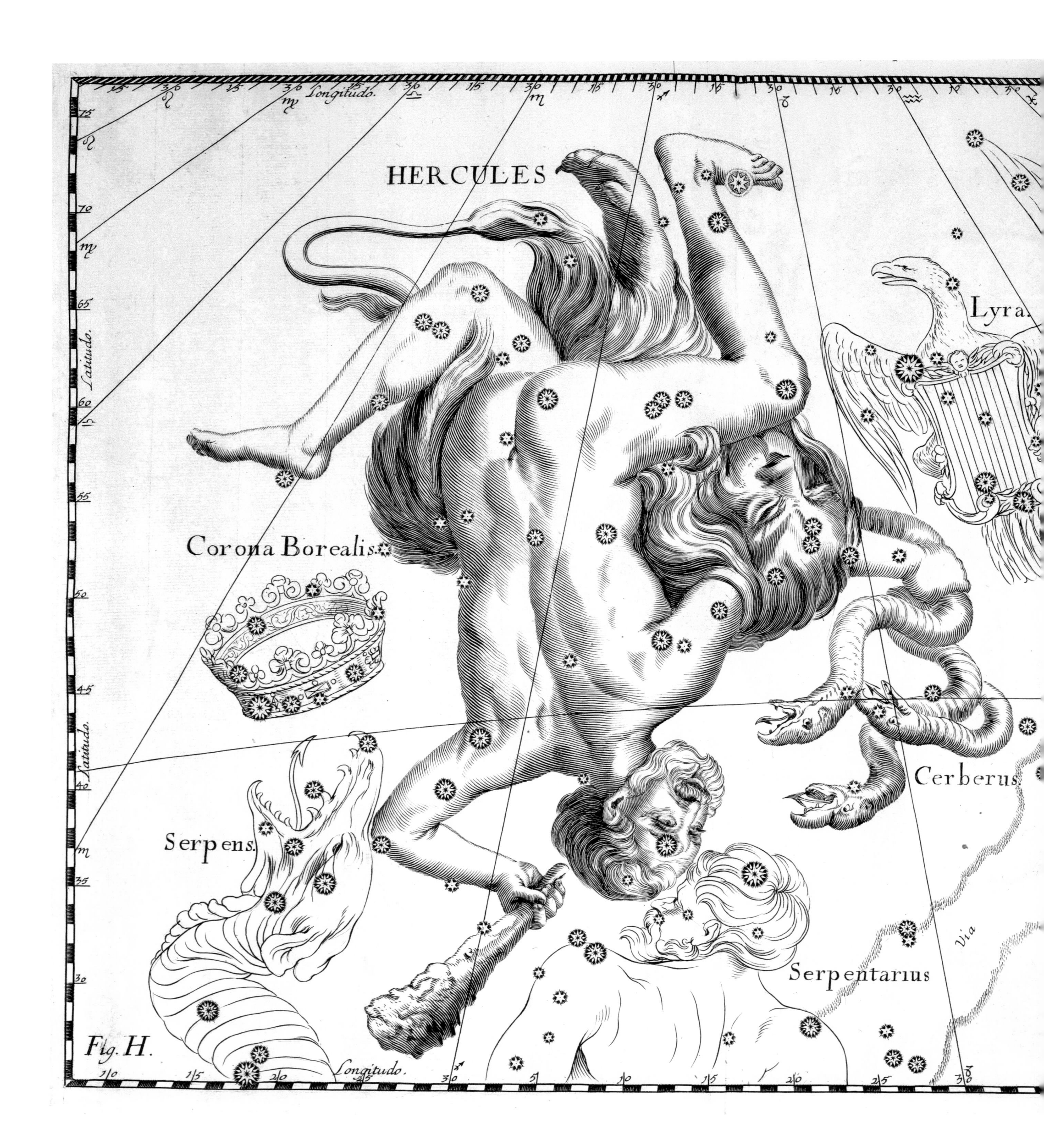
HERCULES
Lyra
Corona Borealis
Cerberus
Serpens
Serpentarius
Via
Fig. H.
Longitudo
Latitudo

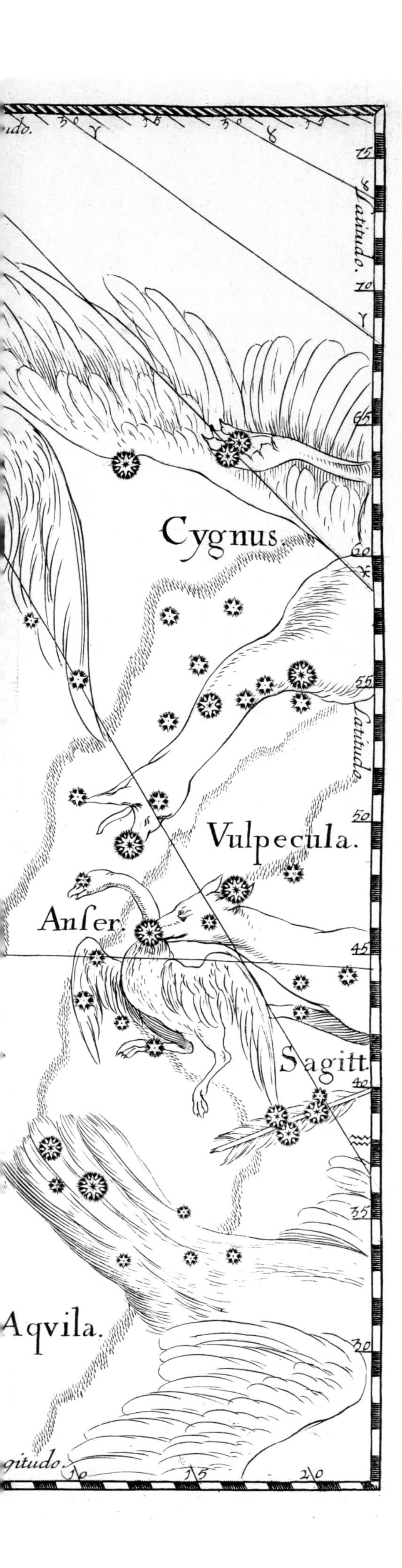

Hercules

Here Hevelius represents the Hercules constellation according to traditional iconography: the mythical hero is armed with a club, and wrapped around his left arm is the pelt of the Nemean lion, which Hercules slew in the first of his Twelve Labors. With the same arm the demigod is strangling Cerberus, the three-headed monster that guards Hades and was killed by Hercules in his last, most difficult and dangerous Labor. The Cerberus constellation was introduced by Hevelius to replace the Ramus Pomifer (Apple Tree Branch) and in order to complete the representation of the hero with greater coherence. However, the author makes use of 'poetic license', departing from classical mythology and depicting the monster as a three-headed snake rather than a three-headed dog.

John Flamsteed

(1646-1719)

A man of intractable character who was hardly inclined to make compromises, the English astronomer John Flamsteed is known for having been the founder of the Greenwich Observatory as well as for his epic clashes with the great Isaac Newton, who also had a rather bad-tempered and unpleasant character. Born in Denby, in Derbyshire, on 19 August 1646, Flamsteed was the son of a maltster who supplied the district breweries with malt. Due to his poor health, the young Flamsteed was forced to interrupt his studies when he was only fourteen. He then concentrated on learning astronomy by himself until he finally succeeded, in 1670, in being admitted to the Jesus College of Cambridge to attend lessons given by Newton, who at the time was already a renowned mathematics professor. The turning point came about in 1675, when he accompanied his patron Jonas Moore to the court of King Charles II and was able to meet the monarch, who was quite impressed by him and appointed him Astronomer Royal, charged with studying the positions of the stars in order make navigation safer. Flamsteed accepted, but stated that this task would be possible only by being able to use more accurate star maps, which in turn could be made only by having an adequate observatory at one's disposal, which in fact did not exist. So Charles II ordered that one be constructed for Flamsteed at Greenwich and even granted his court astronomer the honor of laying the foundation stone.

Flamsteed became director of the observatory with a salary of 100 pounds (from which were deducted the taxes owed and all expenses for the purchase and maintenance of all necessary equipment) and for forty years he devoted all his energy to studying the stars and comets with the aim of writing a long and exhaustive catalog that would be the crowning achievement of a life's work. In truth, his extreme scrupulousness and obstinacy prevented him from publishing the results in installments without their being obsessively verified, since he was haunted by the fear of making a terrible blunder. This behavior exasperated his colleagues, especially Newton, with whom he was now collaborating; both were fellows of the exclusive Royal Society, whose members were the cream of British science. In 1712, needing to consult Flamsteed's highly accurate calculations in order to put the finishing touches to his own, Newton asked for help from his colleague Edmond Halley, with whom he had a cordial relationship, in order to get hold of the draft of Flamsteed's catalog. But the two overdid it, going so far as to publish a controversial edition of the catalog, with the title Historia Coelestis Britannica, *without Flamsteed's authorization. The enraged Flamsteed reacted by immediately cutting all ties with Halley and giving vent to his ill feeling toward Newton. He then went about searching for the copies in circulation, managing to find most of them, which he burned in theatrical fashion right in front of the Observatory. The 'official' version of the* Historia *would be published posthumously by Flamsteed's widow in 1725, six years after his death (which occurred at Greenwich on 12 January 1719), with the title* Stellarum inerrantium Catalogus Britannicus *(now known as the* British Catalogue*). This work was followed, in 1729, by the* Celestial Atlas *or* Atlas Coelestis, *a monumental illustrated atlas that, besides containing almost 3000 stars, set out to correct and improve upon the illustrations of the constellations published by Johann Bayer by restoring the centuries-old usage of depicting the constellations frontally, thus making for more accuracy and more immediate understanding.*

Among Flamsteed's major contributions, mention must be made of the introduction of a new type of nomenclature, as opposed to the one conceived a century earlier by Bayer: he replaced the Greek letters with numbers to classify stars. For example, Flamsteed was the first to catalog the planet Uranus (although he mistakenly indicated it as a star) and did so as 34 Tauri.

Portrait of John Flamsteed in an engraving by George Vertue (1721) based on the original oil painting by Thomas Gibson.

T. Gibson pinx. 1712.
Geo. Vertue Sculp. 1721
JOHANNES FLAMSTEEDIUS Derbiensis
Astronomiæ Professor Regius. Anno Ætatis 74 Obijt
Decem: 31 1719

Atlas Coelestis

(1729)

The *Atlas Coelestis* was published in London in 1729, ten years after Flamsteed's death, by the executors of his will, including his widow Margaret. With its almost 3000 stars, which are double the number listed by Johannes Hevelius and triple those catalogued by Tycho Brahe, this work was the most complete catalog published up to that time. The atlas also contains two planispheres and 26 maps of the principal constellations visible from Greenwich: the mythological representations were executed by the painter James Thornhill.

Flamsteed's work had an immediate positive impact due to its accuracy, the result of his precise observations. Nonetheless, its almost monumental size – the plates are 23.6 x 19.7 inches (69 x 50 cm), which makes it the largest celestial atlas ever published – and its concomitant high price made it difficult for those who needed to consult it. Consequently, in 1776 Jean Nicolas Fortin, a Frenchman specialized in the construction of globes and scientific instruments, published a smaller sized version. No one knows who commissioned this work (perhaps it was one of his scientist customers, or even the French royal family), but the new edition of Flamsteed's atlas, printed in French and a third the size of the original, enjoyed a huge success among the cultured circles of the time. The *Atlas Fortin-Flamsteed*, as it was called then, became a major reference point in the literature of this discipline and was reprinted and updated several times, closely following the progress made in astronomical knowledge step by step.

Celestial Map of the Northern Hemisphere

This beautiful map represents the constellations in the Northern Hemisphere. The plates in Flamsteed's atlas contain many more stars than those in preceding atlases, and also feature celestial bodies that had never before been illustrated. A case in point is the small star that appears in the plate dedicated to Andromeda. This is the M31 galaxy, the first non-stellar object to be placed in a celestial map of this kind.

P. 26.
Monoceros
Orion
Procyon
Canis Minor
Hydra
Gemini
Cancer
Pollux
Castor
Taurus
Pleiades
Cetus
Aries
Auriga
Capella
Lynx
Leo
Leo Minor
Perseus
Triangulum
Camelopardalus
Pisces
Cassiopea
Ursa Major
Andromeda
Comæ Berenices
Ursa Minor
Canes Venatici
Cepheus
Lacerta
Draco
Markab
Pegasus
Arcturus
Bootes
Lyra
Cygnus
Corona
Hercules
Serpens
Cerberus
Sagitta
Equuleus
Delphinus
Aquila
Antinous
Ophiuchus
Magnitudines

Celestial Map of the Southern Hemisphere

Like the preceding plate of the Northern Hemisphere, the one illustrating the Southern Hemisphere also has many more stars than the other plates published in earlier atlases. As the astronomer Felice Stoppa explains, in Flamsteed's work for the first time the stars are represented "through their equatorial coordinates: right ascension and declination, the graticule of which is, in the plates, placed over the polar ecliptic one." In the lower right section of the plate is a legend that explains the magnitudes. Both plates of the hemispheres were republished in 1820 by Richard Brookes, who used the original copper plates engraved by Abraham Sharp but made a serious mistake: the Northern Hemisphere is indicated as the southern and vice versa.

Ophiuchus et Serpens

*According to mythology, Ophiuchus, associated with the Greek god of medicine Asclepius and also known as the Serpent Bearer (*Serpentarius *in Latin), was the son of Coronis (or Arisnoe in other versions), whom Apollo courted and made pregnant. Coronis then betrayed the god with a human being, and when a white crow told Apollo about this, the furious god made the bird's feathers turn black and then took revenge on his lover by killing her with an arrow. However, the god saved his unborn son in Coronis's womb, and entrusted it to the care of the centaur Chiron. One day a serpent told Ophiuchus about a miraculous herb able to raise the dead and became a great healer. Here Flamsteed depicts Ophiuchus while he is 'bearing' the Serpent constellation, the only one that is 'split' into two parts.*

Taurus

The zodiacal constellation of Taurus is one of the most important and well known in the sky. In this plate, Flamsteed represents the powerful Bull charging at the nearby Orion, who defends himself with a club. Corresponding to the animal's right eye is Aldebaran, the brightest star in the constellation and the fourteenth most luminous in the night sky. The Pleiades are also part of the constellation; they rise in the sky just before Aldebaran, which thus seems to be pursuing them (hence the name of the star, which in Arabic means 'the follower'). This plate also introduces 34 Tauri, which Flamsteed mistook as a star but was really the planet Uranus.

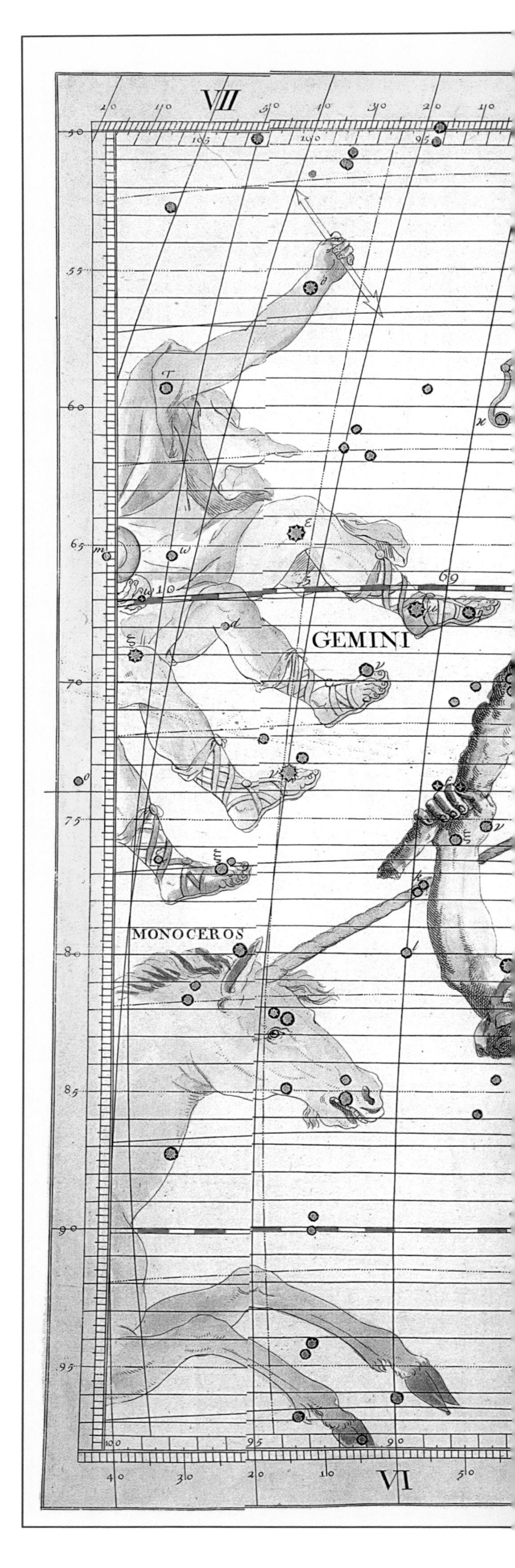

TAURUS
PERSEUS
Algol
TAURUS
Pleiades
ECLIPTICA
Aldebaran
ARIES
RION
CETUS
ERIDANUS
Rigel
IGA

Monoceros, Canis Major et Minor, Navis, Lepus

This plate represents various neighboring constellations: Monoceros or Unicorn, Canis Major and Minor (Greater and Lesser Dog), Navis (Ship) and Lepus (Hare). Unicorn is a modern constellation, therefore unknown to Ptolemy. It was introduced by Petrus Plancius in 1613 and mapped for the first time by Jakob Bartsch in 1624, although some scholars state they have seen it in certain Persian maps. The unicorn is not an animal that belongs to classical mythology, but was well known in medieval bestiaries as a symbol of chastity. On the other hand the two dog constellations were well known already by the ancients, who set them in relation with nearby Orion, as they were his hunting dogs. The Canis Major follows the Lepus constellation and is recognizable because it has the brightest star in the night sky, Sirius.

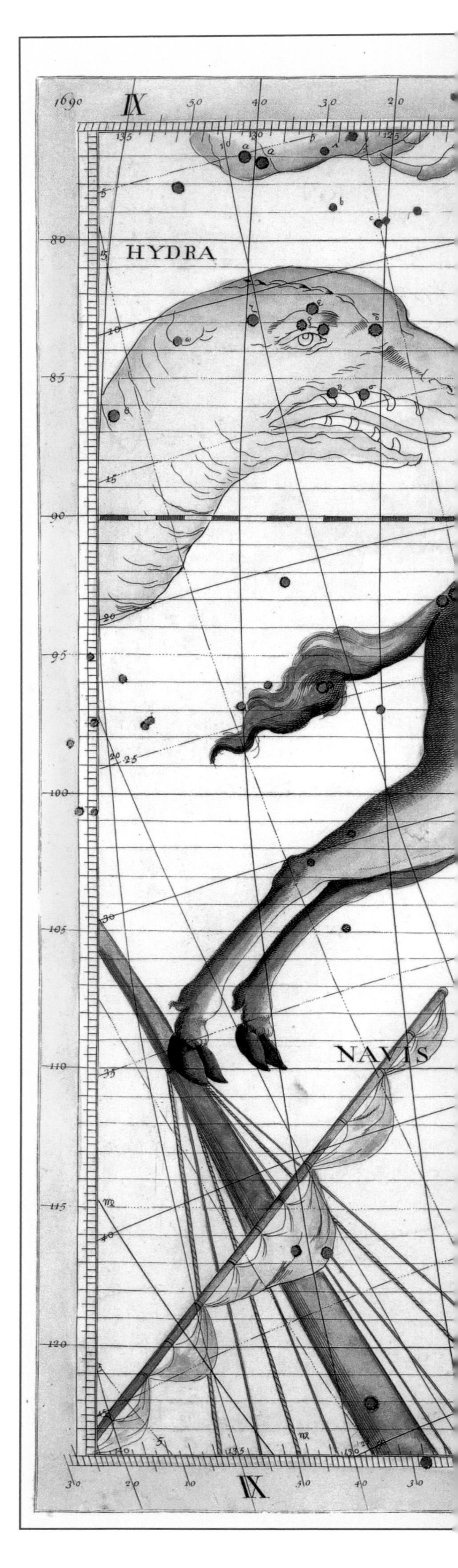

MONOCEROS, CANIS MAJOR & MINOR, NAVIS, LEPUS.
P 13.
CANCER
CANIS
MINOR
ORION
MONOCEROS
Syrius
LEPUS
ANIS MAJOR
COLUMBA

Comae Berenices, Boötes, Canes Venatici

*Besides Boötes and his hunting dogs (*Canes venatici*), this plate illustrates Comae Berenices (Berenice's Hair). This is a group of stars that was already known to the ancients, but they did not classify it as a constellation because they thought is was part of the Leo constellation. Ptolemy described it as a "nebulous mass" that looked like a lock of hair, going back to what Eratosthenes of Cyrene had said, comparing the asterism to the hair of Berenice, the consort of Ptolemy III Evergete, king of Egypt. According to the legend, Berenice made a vow to cut off her hair should her husband return victorious from the war, and so it was. But the tresses were immediately transformed into a group of stars that was placed next to Leo's tail. Comae Berenices became a bona fide constellation in 1551 and in 1602 was included in Tycho Brahe's catalogue.*

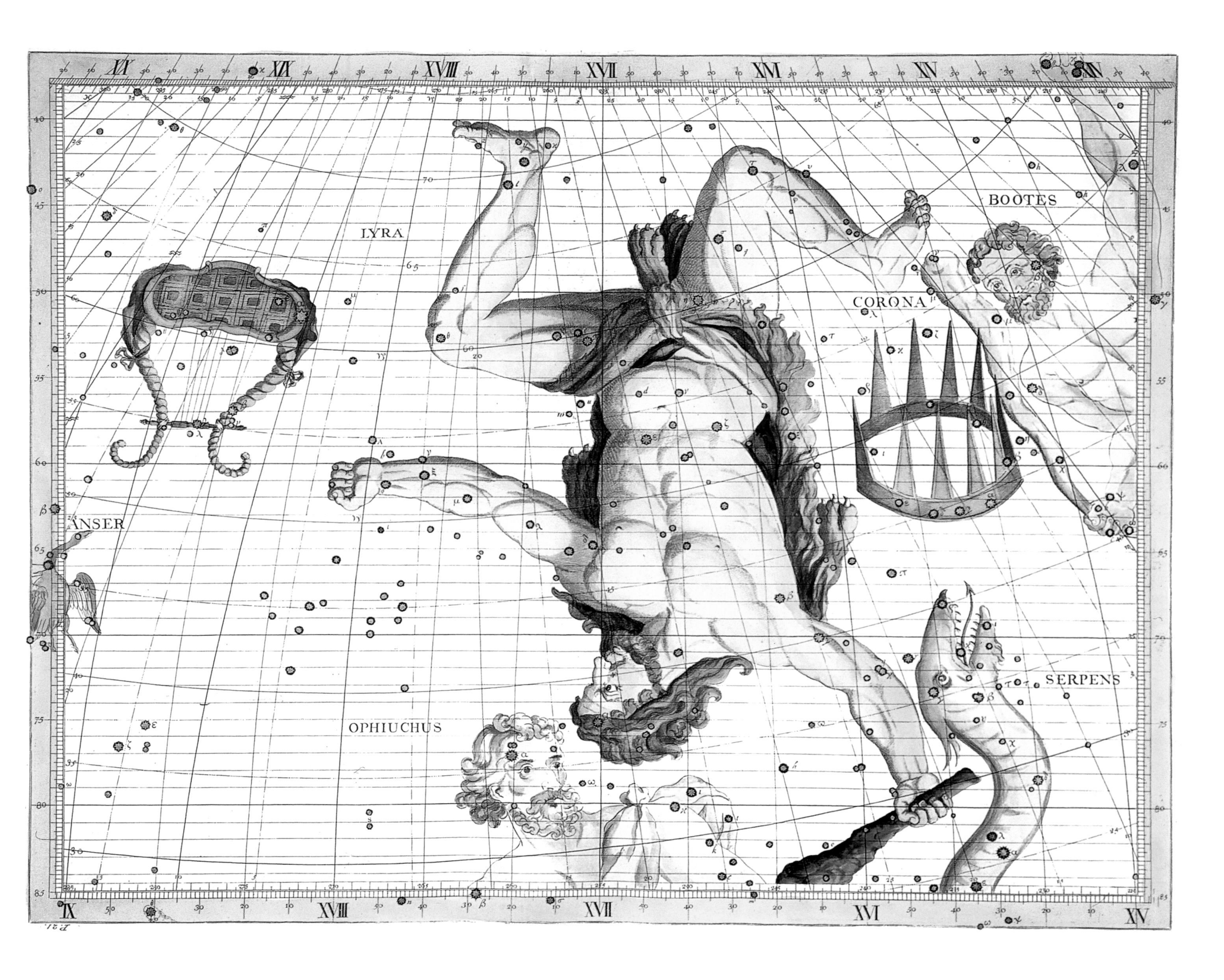

Hercules, Corona et Lyra

Respectively to the right and left of Hercules are the constellations Corona Borealis (Northern Crown) and Lyra (Lyre). In classical mythology this crown (which is not to be confused with the corresponding Southern Hemisphere constellation) is the diadem given to Ariadne, the daughter of Minos, by the god Dionysus, who married her after she was abandoned by Theseus, the hero who killed the dreaded Minotaur. Lyra on the other hand is associated with Orpheus's lyre, which the god Hermes made from a tortoise shell he had found in Arcadia, adding seven cow gut strings, which is the same number as the Pleiades (however, here Flamsteed depicts only five). After Orpheus died and joined his beloved Eurydice in Hades, the instrument became a constellation, which has the fifth brightest star in the sky, Vega (Alpha Lirae), while the second brightest is the star system Beta Lyrae or Sheliak, which means 'harp' in Arabic.

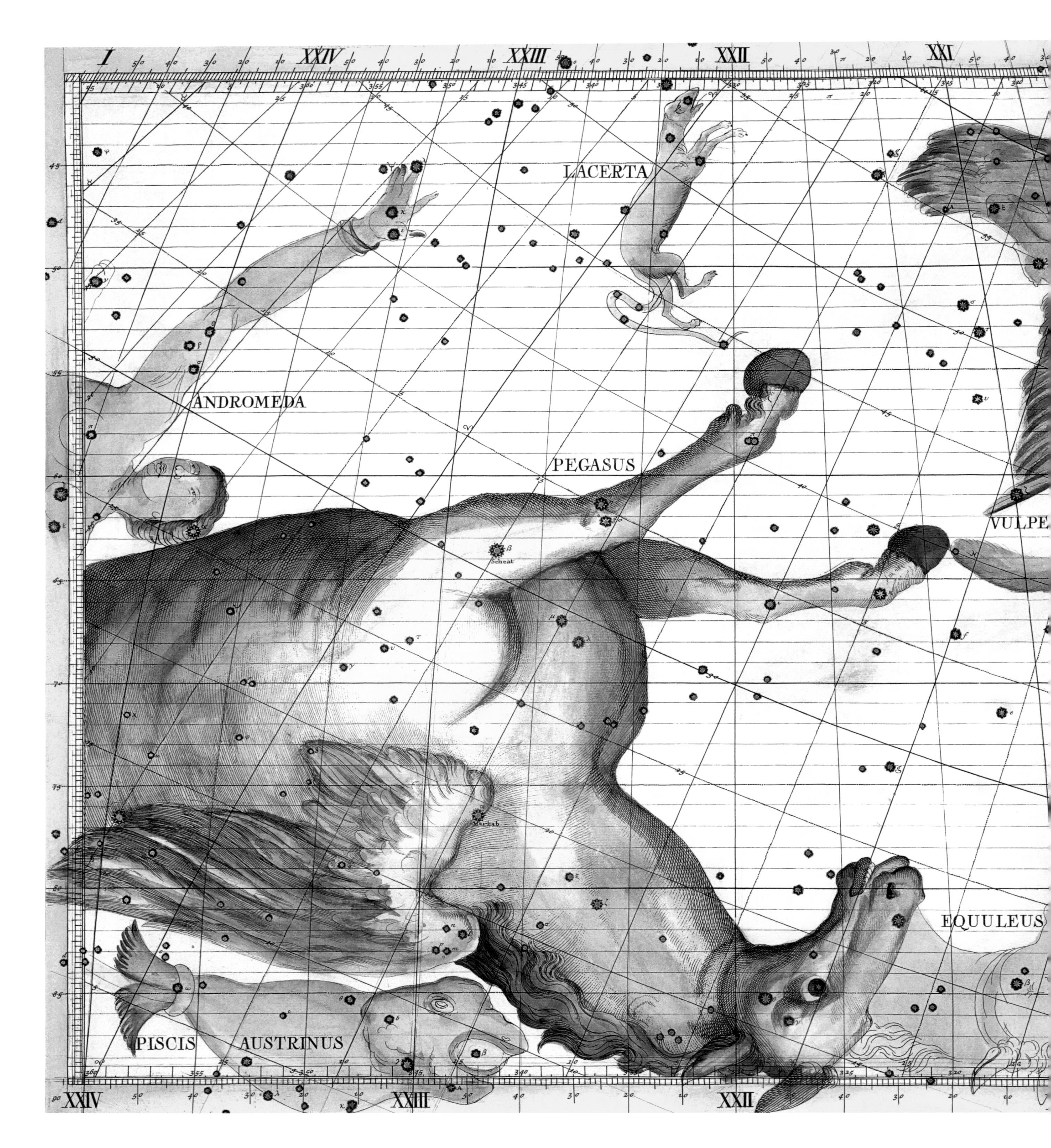
LACERTA
ANDROMEDA
PEGASUS
Scheat
Markab
VULPE
EQUULEUS
PISCIS AUSTRINUS

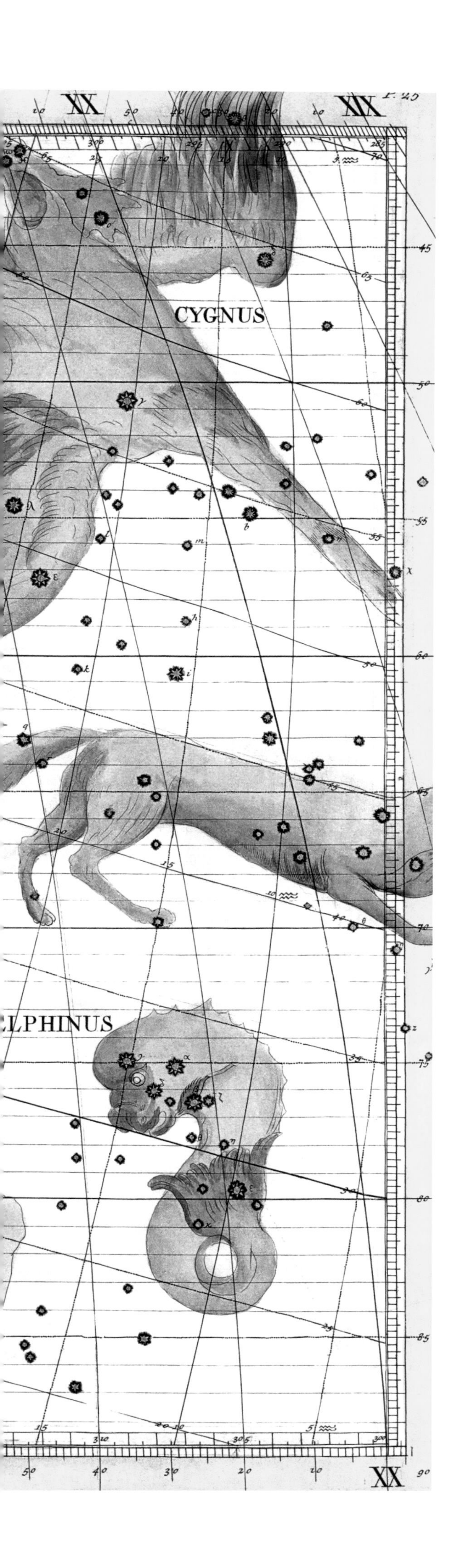

Pegasus et Equuleus

The Pegasus constellation is named after the winged horse of the same name that, according to Greek mythology, was born from the source of the ocean (in fact the name means 'spring') immediately after Perseus had killed the Medusa. It is close to Andromeda, the daughter of Cepheus whom Perseus saved from the sea monster. Pegasus was caught by the hero Bellerophon, who tamed it and rode it to perform many exploits. After the death of the hero, Pegasus was transformed into a constellation by Zeus. Next to him is another constellation, Equuleus or Little Horse, which has no bright stars and is therefore rather faint. It is associated with the foal Celeris, the brother of Pegasus, which the god Hermes/Mercury gave to Castor. Strangely enough, Flamsteed chose to depict only the head, trunk and fore hooves of Pegasus.

Johann Gabriel Doppelmayr
(1677-1750)

Johann Gabriel Doppelmayr was born on 27 September 1677 in Nuremberg to a family of well-to-do merchants. His father, who was an enthusiast of applied physics, soon transmitted his passion for scientific disciplines to his son, and after finishing his studies at the local Gymnasium, the young Johann Gabriel enrolled at the University of Altdorf, where he studied jurisprudence, as well as mathematics and physics, graduating in 1698 with a dissertation on the Sun. He then enrolled at the University of Halle to continue these studies. The following two years he traveled to Berlin, Holland and England, where he met leading astronomers and other scientists and academics. During these important travels Doppelmayr also learned French, English and Italian and perfected his knowledge of several disciplines, including physics and astronomy, and also mastered other skills, such as grinding lens and making telescopes. He finally returned to his native town, Nuremberg, and in 1704 was appointed professor of Mathematics at the Aegidien-Gymnasium, where he had studied in his youth. He would stay at this post for the rest of his life, publishing articles, books and dissertations on various subjects, even including a sort of compilation of short biographies of all the mathematicians, scientists and makers of scientific instruments from Nuremberg.

In February 1716 he married Susanna Maria Kellner; they had four children, three of whom died shortly after their birth. He managed to overcome his grief through his passion for study, to which he devoted all his energy, becoming increasingly more interested in astronomy and celestial cartography. The decisive meeting in this regard was the one with Johann Baptist Homann, a former Dominican monk of Swabian origin who converted to Protestantism and moved to Nuremberg in 1668, working as an engraver for specialist cartography publishers.

In 1702 Homann founded his own publishing firm, which soon played a major role in this field thanks to the privileges granted to him by Emperor Charles VI, who named him Imperial Geographer. Homann's Grosser Atlas üeber die ganze Welt *(Great World Atlas), published in 1716, is an absolute masterpiece of its kind and a milestone, thanks to the idea, introduced for the first time, of painting in the same color all the territories belonging to same monarch or to a single nation.*

For Homann, Doppelmayr was exactly the right person who could guarantee the precision needed for the astronomical atlases he intended to publish together with the terrestrial maps. The latter therefore prepared the astronomical plates that were to be used in the series of atlases published by Homann (and by his heirs after his death) from 1707 to 1740. For that matter, Doppelmayr had already specialized in cartographical work, partly thanks to the collaboration he had established with the engraver Johann Georg Puschner (1680-1749). From 1728 on the two created three pairs of terrestrial and celestial globes of great beauty and quality. The plates that Doppelmayr made for Homann's single atlases were later collected and published together in 1742, resulting in his most important work, Atlas Coelestis in quo Mundus Spectabilis.

Doppelmayr died in December 1750 at Nuremberg, according to some sources due to the effects of an electric shock he had suffered during an experiment he was carrying out with condensers. However, other sources state that he had carried out those experiments many years earlier, so the accident was in no way connected to his demise. Thus the actual circumstances behind his death are still wrapped in mystery.

Hemisphaerium Coeli Boreale

This plate by Doppelmayr, number 16 in his Atlas Coelestis, *illustrates the northern sky with the constellations that were visible in 1730. At the four corners, couples of cherubs are holding astronomical instruments.*

HEMISPHÆRIUM COELI BOREALE,

quo loca Stellarum fixarum secundum Æquatorem, per Ascensiones nempe rectas et Declinationes ad annum Christi 1730 completum sistuntur

à IOH. GABRIELE DOPPELMAIERO *Mathem. Prof. Publ. Academ. Cæsar. Leopoldino-Carol. Nat. Curios. et Reg Societatis Boruss. Sodali.*

Operâ IOH. BAPT. HOMANNI SAC. CÆS. MAJ. GEOGR. Norimbergæ. *Cum Privilegio S.C.M.*

Atlas Coelestis

(1742)

The original title of the *Atlas Coelestis* is 'astronomical', to say the least: *Atlas Coelestis in quo Mundus Spectabilis et in eodem Stellarum omnium Phoenomena notabilia, circa ipsarum Lumen, Figuram, Faciem, Motum, Eclipses, Occultationes, Transitus, Magnitudines, Distantias, aliaque secundum Nic. Copernici et ex parte Tychonis de Brahe Hipothesin. Nostri intuitu, specialiter, respectu vero ad apparentias planetarum indagatu possibiles e planetis primariis, et e Luna habito, generaliter e celeberrimorum astronomorum observationibus graphice descripta exhibentur, cum tabulis majoribus XXX.* This work was published in 1742 by the heirs of Johann Baptist Homann, who had died in 1724. Although it contains no major scientific discovery, it stands out for the astounding beauty of the charts.

There are thirty plates all told, almost all of which were already published in the individual atlases that Homann had published from 1707 on, with the help of Doppelmayr himself. Of these, some represent the sky in its multifarious aspects, including the motion of the planets and the appearance of the solar system with a description of the solar eclipse observed on 12 May 1706; others illustrate the astronomical theories formulated by Nicolas Copernicus and Tycho Brahe. Important contributions are the maps of the systems of the satellites of Jupiter and Saturn, which were created thanks to the studies conducted in 1661 by Giovanni Domenico Cassini, in Bologna, as is the detailed reproduction of the Moon's surface based on the telescopic observations made by Johannes Hevelius and Giovanni Battista Riccioli, which are here compared schematically. Lastly, there are the representations of the comets and their movement, the star maps, and the constellations in the Northern and Southern Hemispheres, for the notation of which Doppelmayr shows he preferred the numbers used by Johannes Hevelius in *Firmamentum Sobiescianum sive Uranographia* to the Greek letters introduced in 1603 by Johann Bayer.

Frontispiece

Doppelmayr's Atlas Coelestis *has a frontispiece depicting the astronomers Ptolemy, Copernicus, Tycho Brahe and Johannes Kepler in a curious landscape with an exotic touch. In the middle is a celestial globe with the constellations. Two cherubs are holding the representation of the sky, as if were a fabric, conceived as a series of concentric systems. In some hand-colored versions of the frontispiece this part is painted in vivid yellow, orange and red.*

Omnia mutantur, siut
sol et terra movetur.
O curas hominum! ô quan
tum est in rebus inane!
Ioh. Keplerus.
Ne frustra vi-
xisse videar.
Tycho Brahe.
ATLAS COELESTIS
studio et labore
IOH. GABRIELIS DOPPELMAIERI. Math. P.P.
impensis
Heredum Homannianorum
Noribergæ. A. MDCCXLII

Systema Solare et Planetarium

The second plate in Doppelmayr's Atlas is truly spectacular, a summary of early-18th century knowledge of the solar system. The base has the heliocentric theory formulated by Copernicus, with the concentric orbits of the planets (including Earth) up to the wheel of the Zodiac, which 'terminates' the cosmos. In the upper left is a reproduction of the individual planets, whose size is compared to that of the Sun; on the opposite side are the other systems that were thought to gravitate around the stars in the firmament. In the lower left is a description of the solar eclipse that occurred on 12 May 1706 and at right, below a representation of a lunar eclipse, are diagrams of Ptolemy's, Copernicus's and Tycho's systems. Note the elegance with which Doppelmayr shows which of the three he considers valid: the Copernican is the only one held up by Urania.

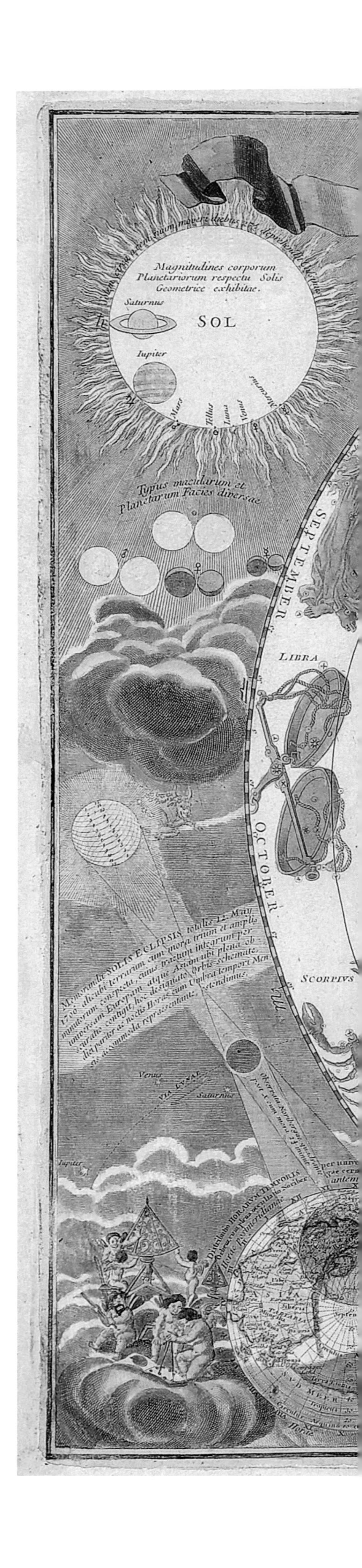

SYSTEMA SOLARE ET PLANETARIVM
ex hypothesi Copernicana secundum elegantissimas Illustrissimi quondam Hugenii deductiones novissime collectum & exhibitum
à IOHANNE BAPT. HOMANNO NORIBERGAE. Cum Privilegio Sac. Caes. Maj.
CANCER
LEO
VIRGO
GEMINI
TAURUS
ARIES
PISCES
AQUARIUS
CAPRICORNUS
SAGITTARIUS
APRILIS
MARTIUS
FEBRUARIUS
IANUARIUS
DECEMBER
ORBITA SATURNI
SYSTEMA SATURNI
ORBITA IOVIS
SYSTEMA IOVIALE
ORBITA MARTIS
ORBITA TERRAE
Aphelium ♃
Perihelium ♃
EX HIS CREATOREM
PROPORTIONES
Benevole Spectator.
TERRA
SYSTEMA TYCHONIS
SYSTEMA COPERNICI

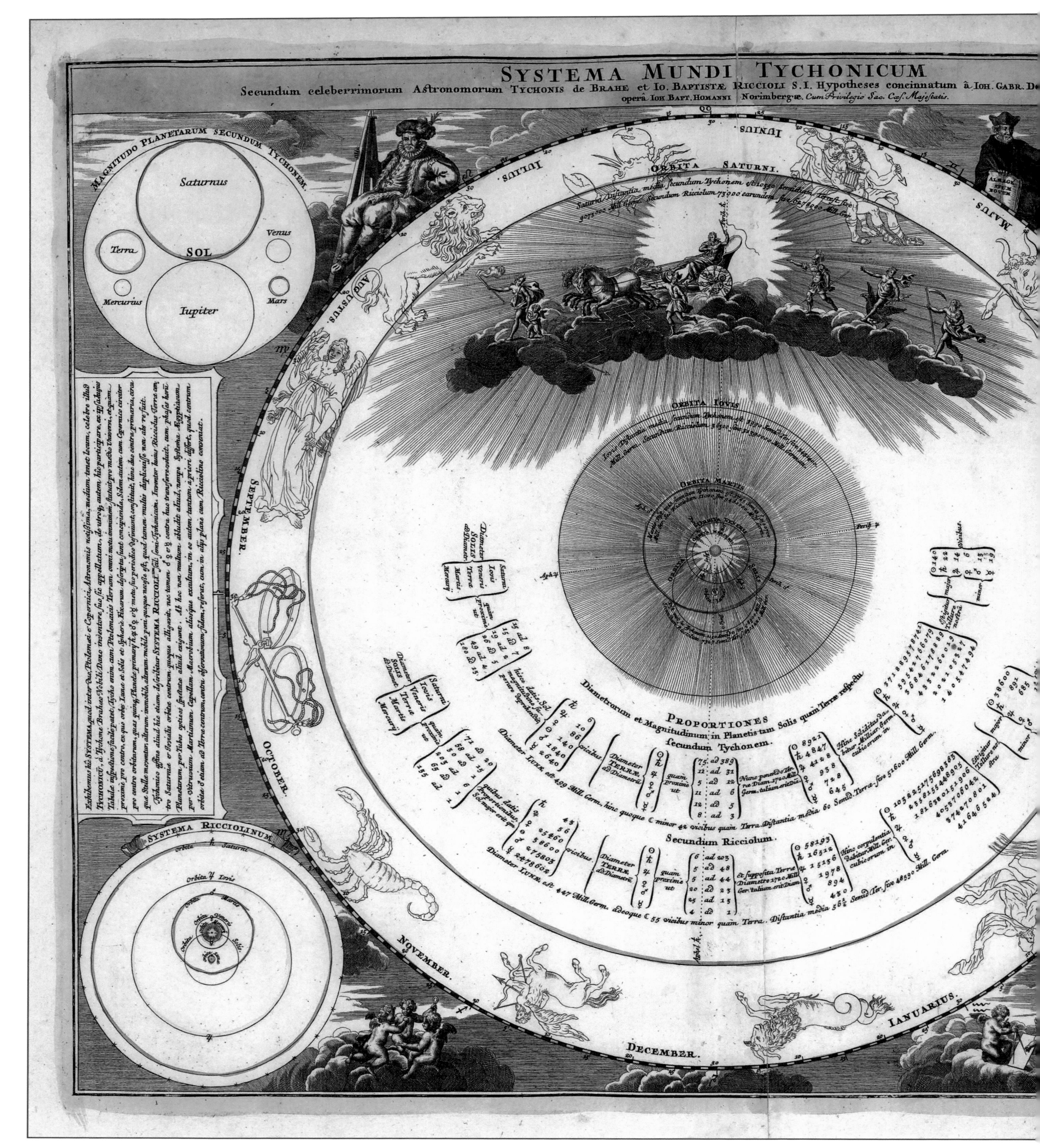
SYSTEMA MUNDI TYCHONICUM
Secundum celeberrimorum Astronomorum TYCHONIS de BRAHE et IO. BAPTISTÆ RICCIOLI S.I. Hypotheses concinnatum à IOH. GABR. D
operâ IOH BAPT. HOMANNI Norimbergæ. Cum Privilegio Sac. Cæs. Majestatis.
MAGNITUDO PLANETARUM SECUNDUM TYCHONEM
Saturnus
Terra
SOL
Venus
Mercurius
Iupiter
Mars
SYSTEMA RICCIOLINUM
ORBITA SATURNI
ORBITA IOVIS
ORBITA MARTIS
PROPORTIONES
Diametrorum et Magnitudinum in Planetis tam Solis quam Terræ respectu
secundum Tychonem.
Secundum Ricciolum.
IUNIUS
IULIUS
MAIUS
AUGUSTUS
SEPTEMBER
OCTOBER
NOVEMBER
DECEMBER
IANUARIUS

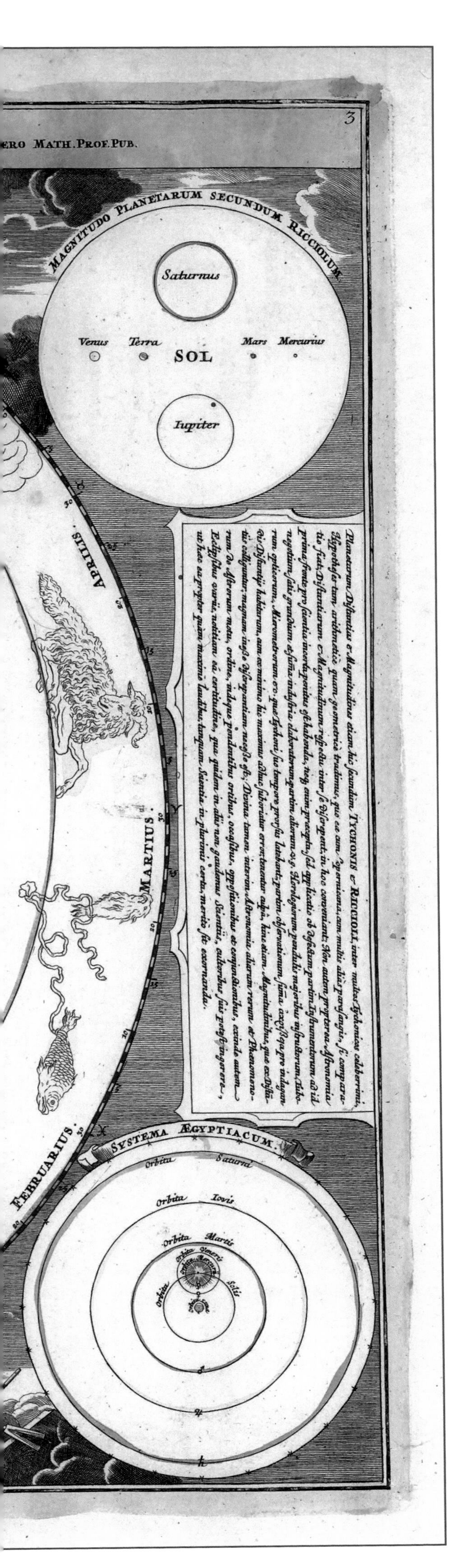

Systema Mundi Tychonicum

This plate, the third of the series, represents a synthesis of the system conceived by Tycho Brahe (thus called Tychonic), which is an intermediate course between the geocentric theory of Ptolemy and the heliocentric one formulated by Copernicus. Above and beyond its illustrative and scientific value, this is certainly one of the most successful and spectacular maps in Doppelmayr's atlas. The details in the illustrations, executed in the usual rococo style, are simply exquisite. For example, besides the ever-present cherubs, who are playing with astronomical instruments, there is the beautiful allegorical representation of the planets in the celestial globe, on which are seated the modern astronomers Brahe (at left) and Riccioli (at right, holding a copy of his Almagestum Novum*).*

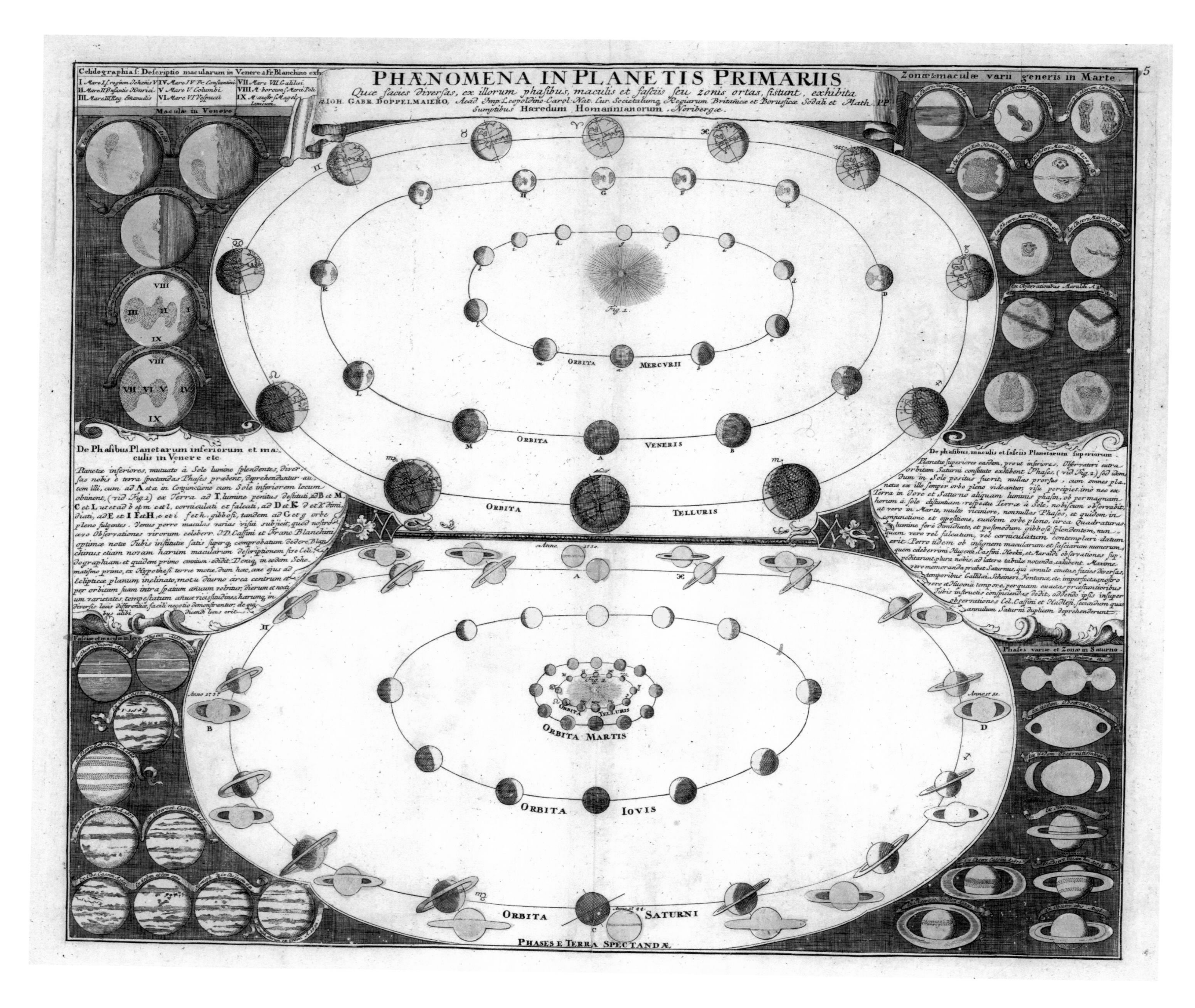

Phaenomena in Planetis Primariis

Here Doppelmayr illustrates the phases of the planets. The information here was gathered from the studies of Christoph Scheiner, Galileo Galilei, Christiaan Huygens, Giovanni Domenico Cassini, Robert Hooke, Jacques-Philippe Maraldi and Francesco Bianchini. This plate, the fifth in the series, contains new data regarding Saturn, which had been observed a short time earlier. The shape of this planet, first viewed in 1610 by Galileo, who considered it 'bizarre', had been sighted thanks to the technical progress made in the construction of telescopes. The rings were registered in 1649 by Eustachio Divini, but their nature remained a mystery until the initial studies made by Huygens (who also discovered this planet's first moon, Titan) led Cassini to explaining it by also observing four other moons of Saturn: Rhea, Iapetus, Dione and Tethys.

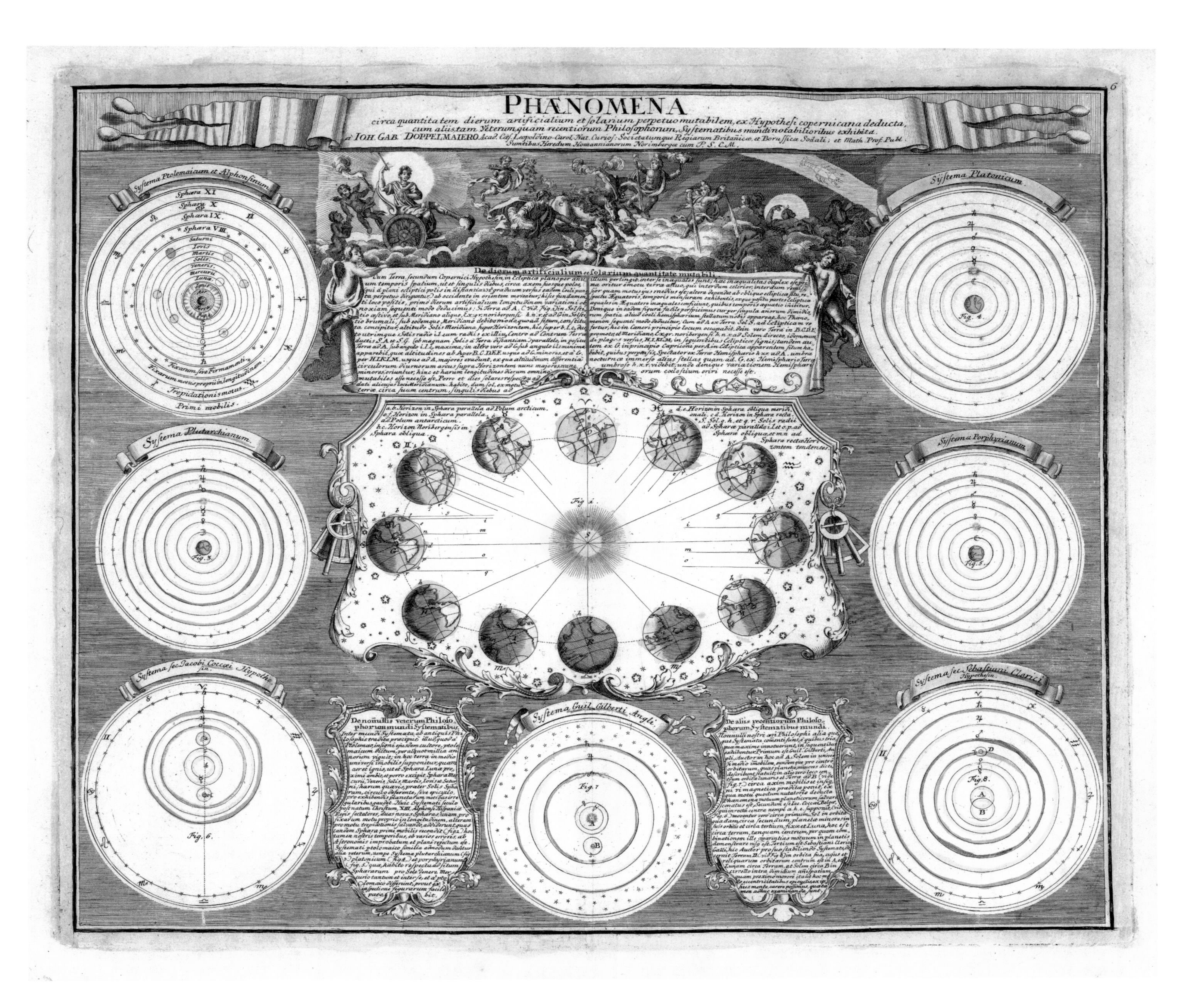

Phaenomena circa Quantitatem Dierum Artificialium et Solarium

In the sixth, truly beautiful plate, Doppelmayr shows us, in the middle, Earth's annual orbit around the Sun framed by seven diagrams that illustrate seven systems formulated by illustrious philosophers and authors of the past (including Plato, Porphry, Plutarch, William Gilbert and Sebastiano Clerico), as well as cartouches with the related explanatory captions. The first diagram at top left, which contains Ptolemy's theories as up-dated in the Alfonsine Tables, was compiled at the behest of King Alfonso X the Wise of Castille and León, who around 1252 entrusted this task to about fifty Arab and Jewish astronomers.

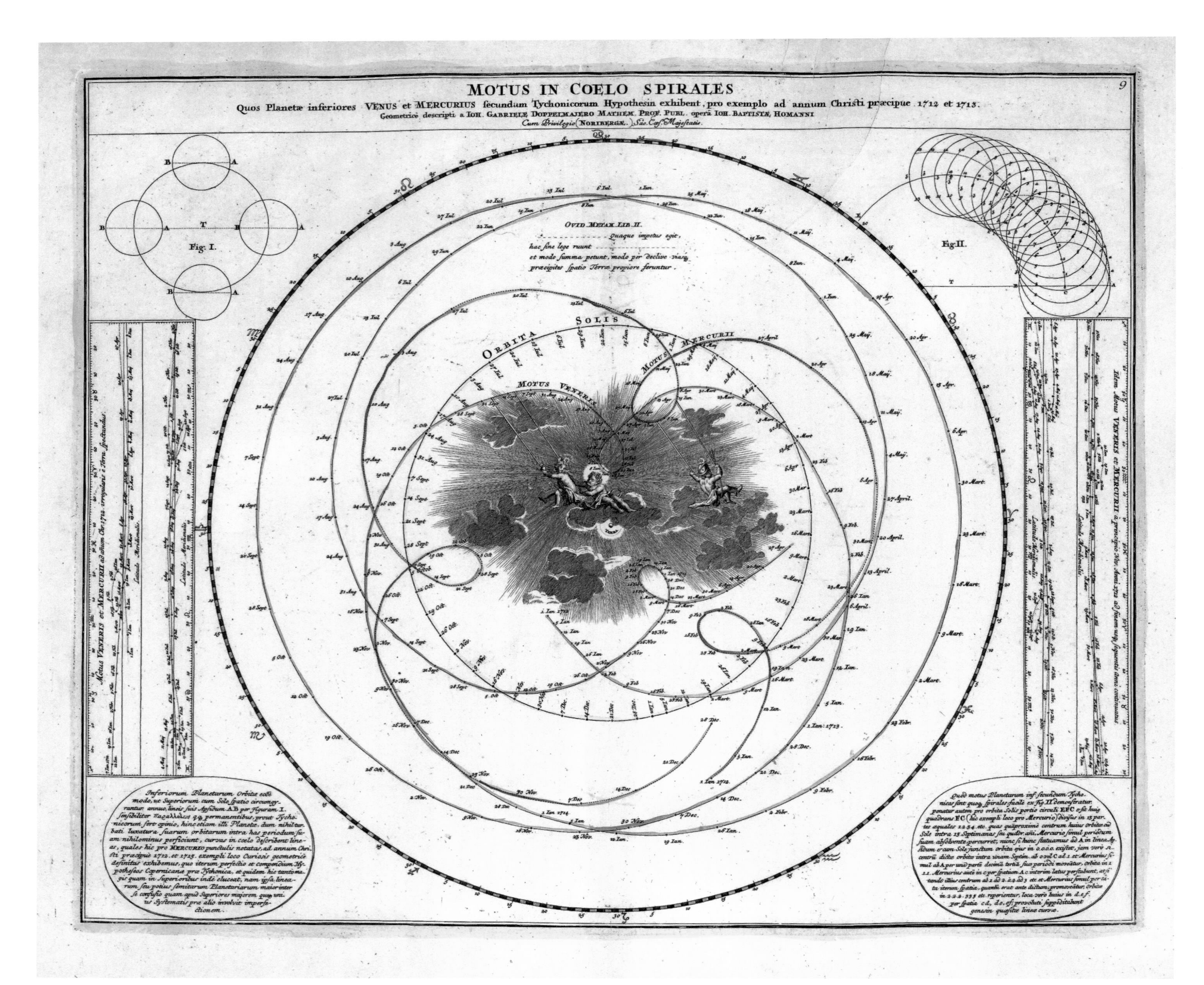

Motus in Coelo Spirales

This table, the ninth of the series, illustrates the predicted geocentric motion of the inner planets Mercury and Venus for the years 1712 and 1713 on the basis of calculations carried out by Tycho Brahe. The dates are marked by means of bars along the course of each of the two planets; in the hand-colored editions these courses are marked with different hues. In the middle, just above the very small terrestrial globe, a fanciful scene with cherubs having fun on swings imparts a carefree and playful aura to the celestial map, in keeping with the rococo taste of the time.

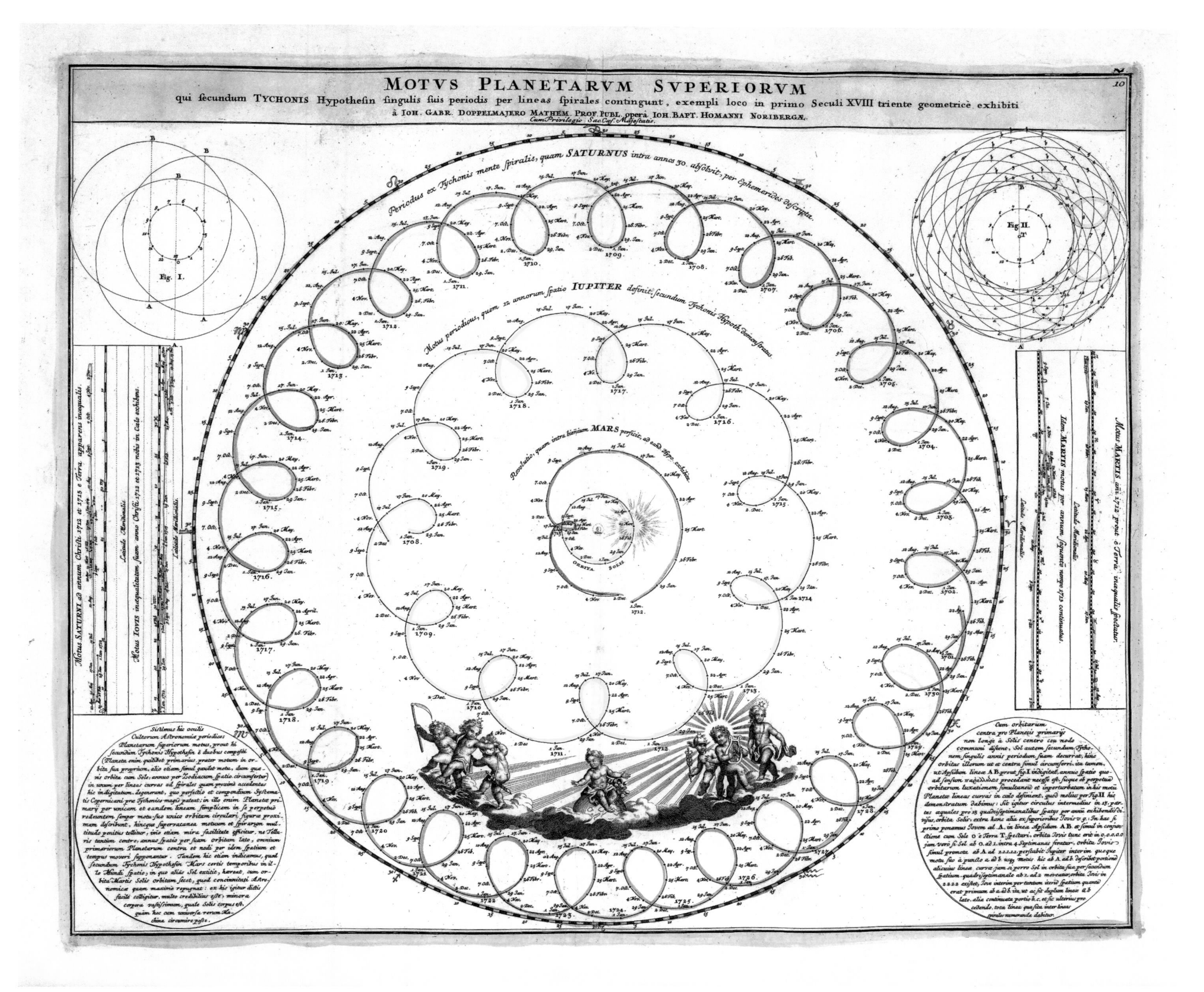

Motus Planetarum Superiorum

The tenth plate is a complement to the fifth one, which describes the motion of the inner planets. In fact, it illustrates the movement of the so-called superior planets, that is, the ones outside the orbit of Earth: Mars, Jupiter and Saturn. The orbits reproduced here, which were reconstructed on the basis of the Tychonic system, correspond to the course the planets followed respectively in 1712 and 1713, from 1708 to 1719, and from 1701 to 1730. Furthermore, in the middle is an indication of the orbita Solis, that is to say, the apparent movement of this star in the sky. The geometric configuration of the map is softened by the delightful representation of the planets in the likeness of cherubs dancing and playing in the sky.

Hemispaerium Coeli Boreale

In the map of the Northern Hemisphere Doppelmayr chooses to 'photograph' the northern sky as it appeared in 1730 and to improve the map with two lists in which he classifies the stars in this hemisphere. The four observatories depicted on the corners are Tycho Brahe's famous Uraniborg, founded in 1576 on the island of Hven but destroyed in the period in which the atlas was published, then the one in Paris (1667), the one in Danzig used by Johannes Hevelius (c. 1650), and lastly Georg Christoph Einmart's observatory in Nuremberg (1678).

ISPHÆRIUM
COELI BOREALE
IOH. GABRIELE DOPPELMAIERO Mathem. P.P. Acad. Imp. Leop. Car. Nat. Curiof. nec non Reg. Scient. Acad. Boruſ. Socio
Sac. Cæs. Maj. Geogra. Norimbergæ.
OBSERVATORIUM PARISIENSE.
STELLÆ HEMISPHÆRII BOREALIS
Dodecatemoria dimidia versus Boream.
OBSERVATORIUM NORIBERGENSE.
Gemini
Pollux
Castor
Taurus
Auriga
Lynx
Ursa Major
Leo Minor
Perseus
Musca
Triang. Minus
Aries
Triangulum Majus
Camelopardalus
Cassiopeia
Pisces Borealis
Linum Bor.
Cepheus
Chara
Asterion
Ursa minor
Andromeda
Linum Austr.
Draco
Lacerta
Stellio
Pisces
Bootes
Cygnus
Lyra
Pegasus
Pisces Austr.
Corona Borealis
TROPICUS
Vulpecula
Serpens
Anser
Delphinus
Aquarius
Hercules
Sagitta
Cerberus
Aquila
ÆQUATOR
Equuleus
Serpentarius
Antinous
Capricornus
Scutum Sobiesci
Sagittarius
Scorpius

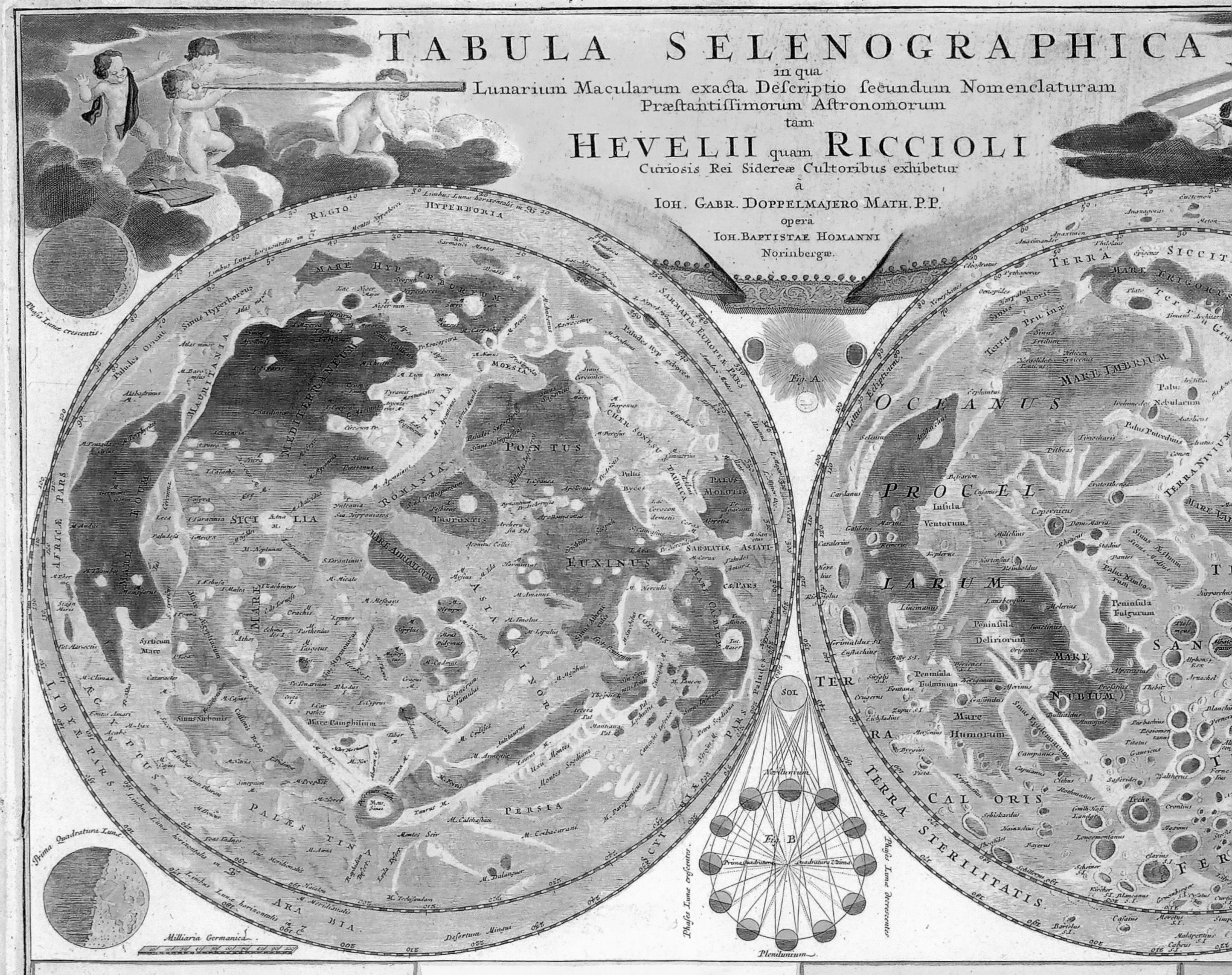

Nullum inter corpora caelestia, ex quo tempore Veteres sacrae Uraniae addicti omnē moverunt lapidem, ut Siderum naturae e affectiones quam maxime forent in aprico positae, cunctorum vicit magis admirationem, e multiformi ambage (si cum Plinio loqui liceat) torsit contemplantium ingenia, proximum quippe ignorari sidus indignantiū, quam ipsa Luna, varietate macularum imprimis miranda; sed nec mirari nos subeat, cum mediis tunc destituti, quibus nunc Lunam accuratius inspicere e contemplari nobis hodie datum, oculis scilicet armatis; hinc etiam deficiente hoc Tuborum opticorum apparatu diversas de Lunae substantia è maculis nudo oculo visis fovere opiniones non potuere non Antiqui illi rei sidereae Cultores; alii enim cum Clearcho e Argesinace maculas Lunares nostri Oceani imaginem in Luna tanquam in speculo conspicuam esse, alii hasce è certis corporibus, quae Solem inter e Lunam jaceant originem ducere existimarunt; alii Lunam vitream, non quidem exacte pellucidā, sed ex parte opacam; alii partim igneam, partim opacam putarunt; e quae sunt multae aliae de corporis Lunaris substantia sententiae.

At multo feliciori successu omnium primus celeberrimus ille Florentinorum Mathematicus Galilaeus de Galilaeis anno superioris seculi decimo, quo utilissimū Tuborum opticorum inventum luci publicae traditum, id negotium tentavit, quod dein Scheinerus e alii satis superq; dedere probatum, imò plures hodie Tubis praedictis ad majorem perfectionem nunc perductis, rem acu, quod ajunt, multo felicius tangere videntur, si pro indubio asserunt, quod Luna innumeris scateat montibus, qui nostros altitudine, habito respectu globi Lunaris ad nostrum, sexagies ferè minoris superent; porro quod eaedem profunditates, quae praegrandibus semper in ambitu suo exteriori, plerumque circulari, maenium instar, cinguntur eminentiis, innumeras ferè et multò plures, sed non tantas et tam profundas, quam nostra exhibeat Terra, si huius cavitates suis destituerentur maribus; denique quod partes multae in Luna obscurae, quae sub primo conspectu non apparent profundae, ideoq; pro materia liquida, maribus scilicet multorum forsan judicio censendae, adhibita accuratiori inspectione, teste Viro celeberrimo D.no de la Hire, nihilominus profundae nec tamen liquidae deprehendantur; ut hinc haud pauci cum acutissimo Galilaeo Lunam pro corpore, materiam à Terra diversam habente existimare possent, in qua etiam fortasse substantiae e res creatae existant, quae operationes edant ab imaginatione nostra sicut remotas, ita e prorsus alienas; quippe quae nullam cum nostris similitudinem habeant, e proin omnino sint à nostra cogitatione discrepantes.

Quamvis autem Luna profunditatibus et eminentiis quamplurimis sit referta, saepissime tamen contingit, superficiem Lunae in certis à Sole distantiis adeo immutatum videri, ut magnus ille montium e profunditatum numerus, qui nuper admodum distinctissime observari poterat, non amplius tunc sub conspectum cadat; ratio huius mutationis ex ipsa figura superiori A intermedia facile patescit, quod scilicet profunditates inter novilunia et quadraturas, Luna crescente à dextris maximè, decrescente autem hac à sinistris potissimum altissimorum circumjacentium montium obtegantur umbris; et quidem quod insuper tales pro vario Solis ad Lunam situ perpetuo immutentur (quae proinde etiam novae maculae denominari solent) eminentiæ autem, cum Sol illas à latere illuminat, quam maximè conspicuæ reddantur; cum è contrario à quadraturis ad oppositionem superficies Lunæ, dum Sol hasce inæqualitatibus magis magisque verticaliter imminere pergit, et omne quidquid umbrosum ante fuit, pedetentim illuminat, aliam semper exhibeat faciem, ut tandem luminosa et albicans appareat.

Ex hoc fundamento bina nostra Schemata in delineatione macularum notabilem etiam differentiam incurrunt, eò quod primum, HEVELIANUM puta Lunâ in oppositione cum Sole existente, hoc est in plenilunio designatum, alterum verò, RICCIOLINUM scilicet, è pluribus Lunæ phasibus in unum corpus fuerit collectum. In denominationibus macularum, utpote signis et significationibus arbitrariis, dictos Auctores inter se differre hic in aperto videmus, cum Hevelius nomina marium, regionum, fluminum et montium nostrorum imitatus, Ricciolus autem illustrium et de re siderea optimè meritorum Astronomorum, complurium præsertim suæ Societatis Mathematicorum nomina pro usu Astronomico sibi elegerit.

Bini circa Lunam limbi se invicem secantes nihil aliud quam motus alicuius in Luna librationis terminos, intra quos perpetua deprehenditur librationis variatio, subindicant; qui hodie demum per Tubos è diversa macularum novilunium mutatione observatus, nec Veteribus olim notus fuit: eandem quippe nobis faciem constantissimè semper Lunam obvertere existimantibus; peragit autem hic motum suum librationis per quatuordecim circiter dies trigesima sexta tantum

Tabula Selenographica

This plate, number 11 in the atlas, features the surface of the visible side of the Moon with an indication of the spots, according to the lunar maps drawn up by Johannes Hevelius (left) and Giovanni Battista Riccioli (right) based on their observations. Of the latter – a Jesuit from Ferrara, Italy and author of the Almagestum Novum *(published in Bologna in 1651) – Doppelmayr provides the nomenclature of the lunar craters, which were dedicated to the leading ancient and modern astronomers. For example, there is the crater Tycho, which Riccioli named in honor of Tycho Brahe, whereas in Hevelius' atlas it is called Mount Sinai. The plate is admirable from an aesthetic standpoint and is completed by a schematic description of the phases of the Moon, made even more charming and elegant by the rococo decoration in the upper section that depicts cherub astronomers and a female allegorical figure, probably Urania.*

Theoria Lunae

Plate number 12 is one of the most complex and schematic in the entire atlas. It represents the movement of the Moon around Earth and its various phases, analogously to what was presented by other, preceding authors in their respective works but utilizing the hypotheses formulated by Isaac Newton. However, the most interesting novelty lies elsewhere: in the upper and lower boxes in the middle of the plate Doppelmayr depicts the surface of the Moon in beautiful chiaroscuro drawings that clearly highlight the dark spots and the craters, and he quotes the nomenclature proposed by the astronomers who had observed and registered them. In the lower box the author even represents a hurtling meteor, depicted just before impact.

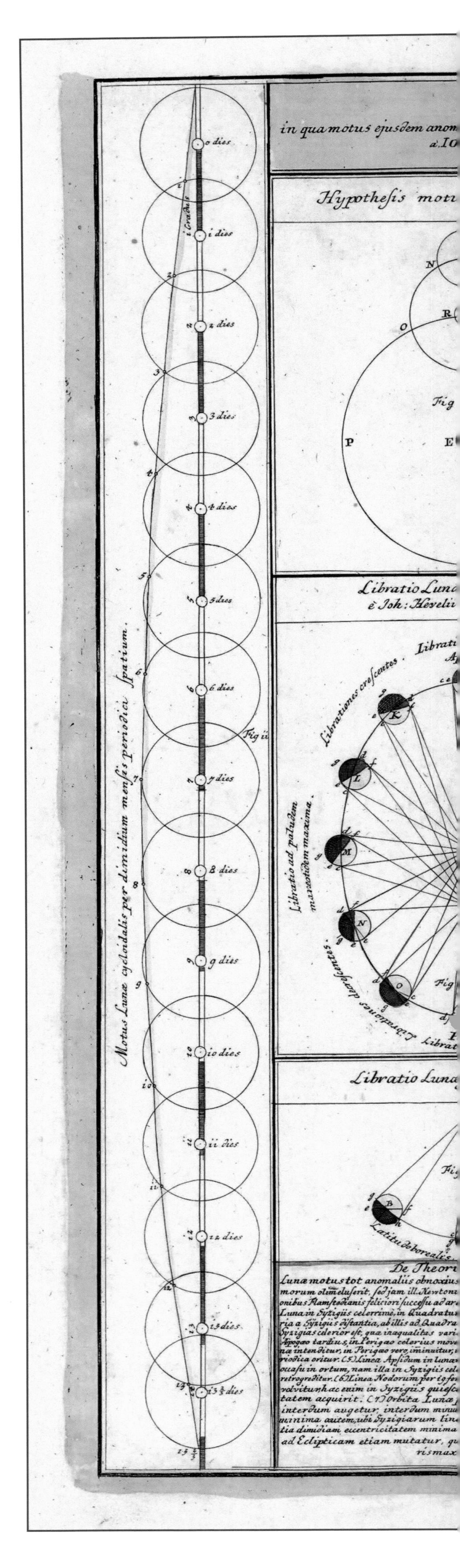

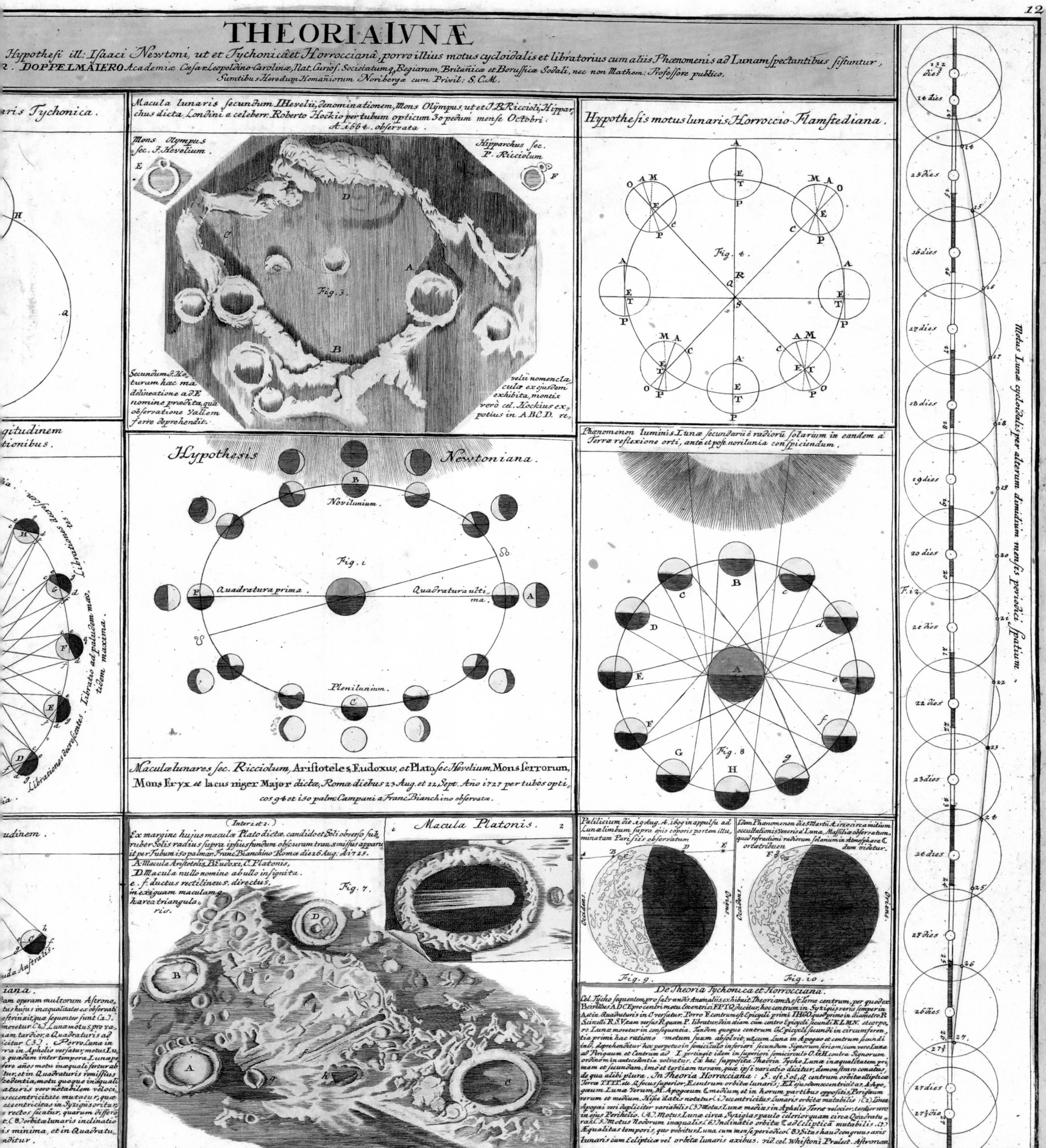
THEORIA LUNÆ
Hypothesi ill: Isaaci Newtoni, ut et Tychonicâ et Horroccianâ, porro illius motus cycloidalis et libratorius cum aliis Phænomenis ad Lunam spectantibus sistuntur,
DOPPELMAIERO Academiæ Cæsar. Leopoldino-Carolinæ, Nat. Curios. Societatumq. Regiarum, Britanicæ et Borussicæ Sodali, nec non Mathem: Professore publico.
Sumtibus Heredum Homannianorum Norimbergæ cum Privil: S. C. M.
Macula lunaris secundum I. Hevelii denominationem, Mons Olympus, ut et I. B. Riccioli Hipparchus dicta, Londini a celeberr. Roberto Hockio per tubum opticum 30 pedum mense Octobri A. 1664. observata.
Mons Olympus sec. J. Hevelium.
Hipparchus sec. P. Ricciolum
Fig. 3.
Hypothesis motus lunaris Horroccio-Flamstediana.
Fig. 4.
Hypothesis Newtoniana.
Novilunium.
Fig. 1
Quadratura prima.
Quadratura ultima.
Plenilunium.
Phænomenon luminis Lunæ secundarii è radiorū solarium in eandem à Terræ reflexione orti, antè et post novilunia conspiciendum.
Fig. 8
Maculæ lunares sec. Ricciolum, Aristoteles, Eudoxus, et Plato sec. Hevelium, Mons Serrorum, Mons Eryx et lacus niger Major dictæ, Romæ diebus 23 Aug. et 22 Sept. Año 1727 per tubos opticos 94 et 150 palm. Campani a Franc. Bianchino observatæ.
Macula Platonis.
Fig. 7.
Fig. 9.
Fig. 10.
De Theoria Tychonica et Horrocciana.
Motus Lunæ cycloidalis per alterum dimidium mensis periodici spatium.
Fig. 12.
Librationes decrescentes.
Librationes crescentes.

THEORIA ECLIPSIUM

in qua variæ Solis occultationes, obscurationes Terræ et Lunæ veræ, stellarum occultationes à Luna, aliaq Phænomena huc spectantia,
a IOH. GABR. DOPPELMAIERO, Acad. Cæsar. Leopoldino. Carol. Nat. Curios. Societatumque Regiarum Britañicæ et Borussicæ, Sodali ut et Mathem
Sumtibus Heredum Homañianorum cum Pr. S. C. M.

Orbis Mercurii sub Sole transitus, qui ab anno 1631 usque ad an. 1736 mense Novembri ad Nodum ascend. (excepto illo a.1661 observati)

Fig. 6.

Týpus eclipsis Lunæ partialis.

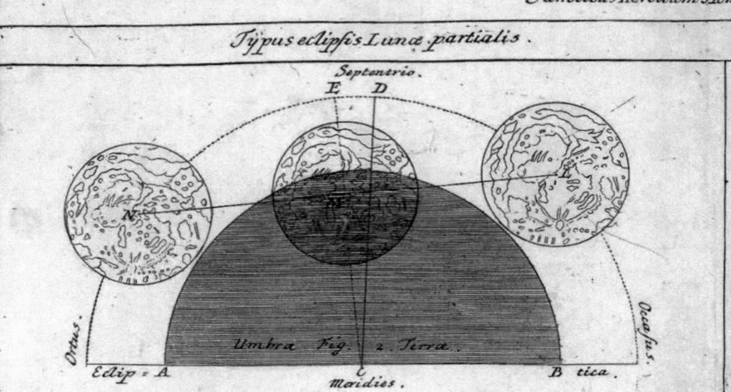

Týpus eclipsis Lunæ totalis cum mora magna et sine m

Tabula, in qua ad. años supra datos dies conjuntionum Mercurii cum Sole, harum Observatores et Observationum loca exhibentur.

Num. Ord.	Anno	Tempus conj. St. n.	Observatores.	Loca Observation.	Num. ord.	Anno	Tempus Conj. St. n.	Observatores.	Loc. Observationum.
1	1631	d. 7 Nov.	Pet: Gassendus.	Parisii.	5	1690	d. 10 Nov.	Wurzelbaur. G. Kirch.	Noriberga. Erfordia.
2	1651	d. 3 Nov.	Jer. Shakerlæus.	Surata in India.	6	1697	d. 3 Nov.	Wurzelbaur. et Cimartus. Cassinus.	Noriberga. Parisii.
3	1661	d. 3 May.	Joh. Hevelius. Hugenius. Streete.	Dantiscum. Londinum.	7	1723	d. 9 Nov.	Edm. Hallejus. Jac. Cassinus	Londinum. Parisii.
4	1677	d. 7 Nov.	Galletius. Edm. Hallejus.	Avenio. Ins. Helena	8	1736	d. 11 Nov.	Manfredius. Marinonius. Chr. Kirchius.	Bononia. Viena. Berolinum.

DIAGRAMMA HIPPARCHICUM.
pro Eclipsibus Solis et Lunæ

Fig. 7.

Graphica designatio orbis veteris fere totius, per cujus maximam partem, et quidem per universam Europam et superiorem Asiæ tractum, Eclipsis Solis (seu potius Terræ) Añi
st. corr. tam totalis quam partialis spectata fuit hic per lineas curvas, illarum indices, exhibita.

Fig. 1.

Loca Terræ in quibus hæc Solis Eclipsis totalis ut plurimum cum mora observata. Terræ loca in quibus eadem Eclipsis p

Fig. 15

Fig. 16.

Týpus eclipseos Solis, seu potius Terræ, universalis.

C. Centrum Terræ.
M.A.I.B.N. Discus Terræ illuminatus.
D.F.H.I.G.E. centrum circuli, intra quem penumbra terminatur.

D.E. via penumbræ.
C.H.I. Linea perpendicularis ad Eclipticam.
C.I.R. linea perpendicularis ad orbitam Lunæ D.E.

Týpus eclipseos Solis partialis in Horizonte quodam.

C. Centrum Solis D.E.
A et B. Centrum Lunæ.
F Centrum Lunæ in medio Eclipseos.

Maculæ Solis insignes a die 9. Novembris usque ad 15 A. 1700 Parisiis observatæ.

Fig. 8.

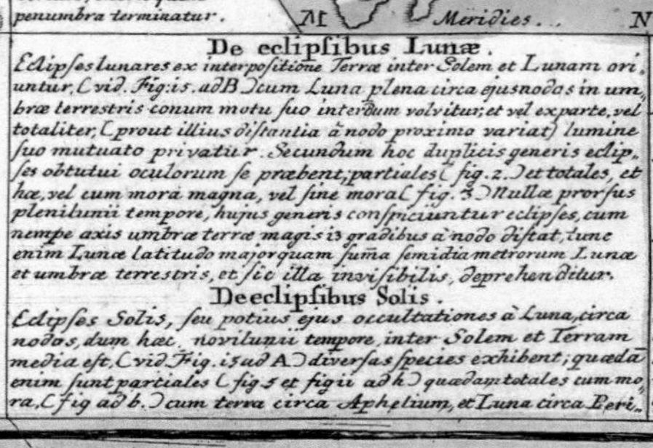

De eclipsibus Lunæ.

Eclipses lunares ex interpositione Terræ inter Solem et Lunam oriuntur, (vid. Fig. 15. ad B) cum Luna plena circa ejus nodos in umbræ terrestris conum motu suo interdum volvitur; et vel ex parte, vel totaliter, (prout illius distantia a nodo proximo variat) lumine suo mutuato privatur. Secundum hoc duplicis generis eclipses obtutui oculorum se præbent; partiales (fig. 2.) et totales, et hæ, vel cum mora magna, vel sine mora (fig. 3) Nullæ prorsus plenilunii tempore, hujus generis conspiciuntur eclipses, cum nempe axis umbræ terræ magis 13 gradibus a nodo distat, tunc enim Lunæ latitudo major quam suma semidiametrorum Lunæ et umbræ terrestris, et sic illa invisibilis, deprehenditur.

De eclipsibus Solis.

Eclipses Solis, seu potius ejus occultationes a Luna, circa nodos, dum hæc novilunii tempore, inter Solem et Terram media est, (vid. Fig. 15 ad A) diversas species exhibent; quædam enim sunt partiales (fig 5 et fig 11 ad h) quædam totales cum mora, (fig ad b.) cum terra circa Aphelium, et Luna circa Perigæum versatur; quædam totales sine mora, (Fig 11 ad g.) cum luminarium diametri apparent æquales, nonullæ vero annulares (fig. 11 ad a) cum Terra in Perihelio et Luna in Apogæo est. Nulla plane Solis eclipsis in Terra apparet, si latitudo Lunæ major existit suma semidiametrorum disci Terræ et penumbræ sive suma semidiametrorum Solis Lunæ et parallaxi Lunæ horizontali, quæ parallaxis æqualis est semidiametro Terræ e Luna spectatæ (vid fig. 7.)

De universalibus Terræ eclipsibus.

In harum eclipsium theoria, secundum quam defectus Solis nobis apparentes, tanquam veræ partium Terræ obscurationes, ex Ioh. Kepleri invento, sistuntur, sequentia notanda. In O. (vid. Fig. 4.) Terræ locus est, in quo umbra lunæ, sole oriente, Terram stringit, et eclipsis incipit; in F. alius est, in quo Sol oriens centraliter tegitur, in H. exacte luminarium conjunctio; in I vero obscuratio maxima contingit, in Terræ loco ad G totalis eclipsis, Sole occidente, cessat, denique in B omnimoda eclipsis, Sole occidente, finitur, hinc totalis eclipseos mora super Terra in tractu F. G; omnimoda autem duratio in tractu D.E. absolvitur. Lineæ 11. 10. 9. 8. &c D.E. ex utraque parte parallelæ, Solis quantitatem, hi in boreali plaga occultati per digitos indicant, qui bus in Eclipsi Solis A. 1706. die 12 Maij totali habi (vid. Fig. 1.) curatius exhibiti sunt.

De minimi generis occultationibus So

Ad hoc genus pertinent Mercurii et Veneris sub Sole tra Mercurii a 1630 ad nostrum usque tempus octo tantum nes nobis iñotuerunt (vid. Fig 6.) Venus hactenus semel a Jerem. Horroccio in Anglia sub Sole observata, (P ro A 1761 observanda. Hisce addimus Solis maculas, ficie, vel in ejus Atmosphæra, nubium aut fumori instar, numero et quantitate sæpius insignes (Fig parte ab Æquatore Solis, intra 30 graduum spatium tubos opticos apparent; moventur illæ motu gyratio trum intra 25½ dies; respectu vero Terræ in orbita motæ, intra 27⅓ dies, quod Phænomenon in Fig 14. exhibetur.

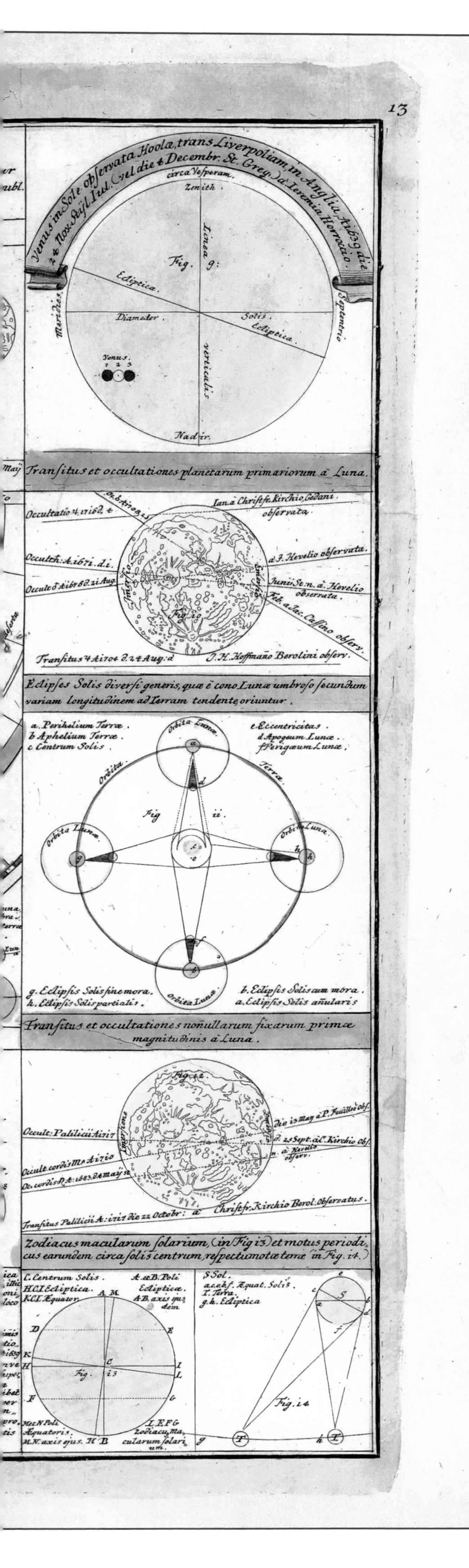

Theoria Eclipsium

Doppelmayr's thirteenth plate illustrates the theory of the eclipses produced by both the Sun and Moon. There are also representations of the various stages of the solar eclipse that occurred on 12 May 1706 as it was seen in Europe and northern Asia. Given their particular nature, phenomena of this type have always been considered exceptionally important since the dawn of human history. Indeed, the ancients viewed them as harbingers of extraordinary events, epoch-making changes and calamities. This was the case in 1706 as well: the eclipse was related to a series of losses suffered by France in the War of the Spanish Succession (Barcelona and Turin) and considered a sign of the imminent end of the power of King Louis XIV, called the Sun King by his contemporaries.

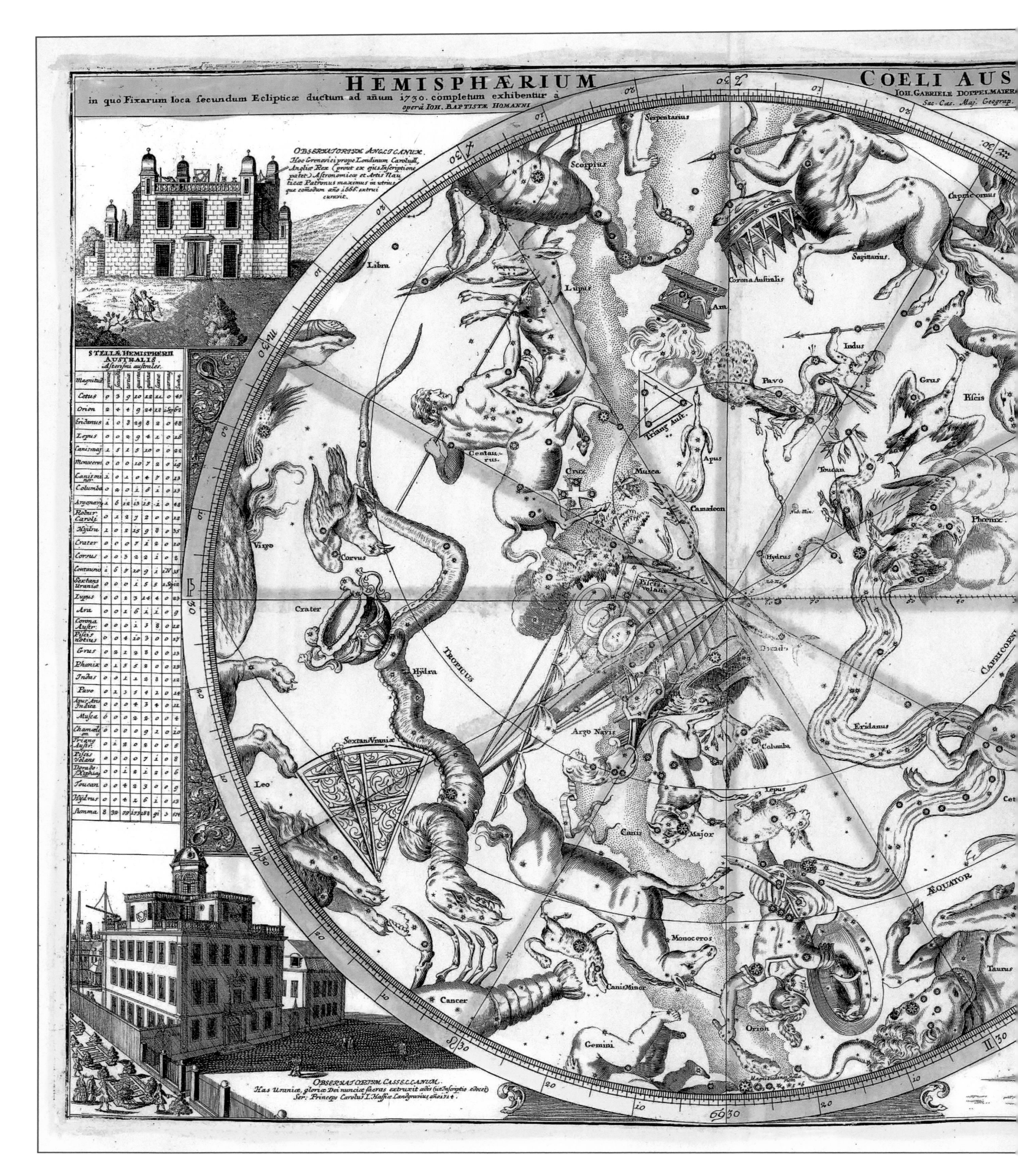
HEMISPHÆRIUM COELI AUS
in quo Fixarum loca secundum Eclipticæ ductum ad añum 1730. completum exhibentur à
IOH. GABRIELE DOPPELMAIERO
operâ IOH. BAPTISTÆ HOMANNI
Sac. Cæs. Maj. Geograp.
OBSERVATORIUM ANGLICANUM.
Hoc Grenovici prope Londinum Carolus II. Angliæ Rex (prout ex ejus Inscriptione patet) Astronomicæ et Artis Nauticæ Patronus maximus in utriusque comodum año 1666. extrui curavit.
STELLÆ HEMISPHÆRII AUSTRALIS. Asterismi australes.
Serpentarius
Scorpius
Sagittarius
Capricornus
Libra
Lupus
Corona Australis
Ara
Indus
Pavo
Grus
Piscis
Triang. Aust.
Apus
Centaurus.
Crux
Musca
Camæleon
Toucan
Phœnix.
Virgo
Corvus
Hydrus
Pisces Volans
Crater
Dorado
TROPICUS
CAPRICORNI
Hydra
Eridanus
Argo Navis
Columba
Sextans Uraniæ
Leo
Lepus
Canis Major
Cet
ÆQUATOR
Monoceros
Canis Minor
Cancer
Orion
Taurus
Gemini
OBSERVATORIUM CASSELLANUM.
Has Uraniæ, gloriæ Dei nunciæ sacras extruxit ædes (ut Inscriptio edocet) Ser: Princeps Carolus I. Hassiæ Landgravius año 1714.

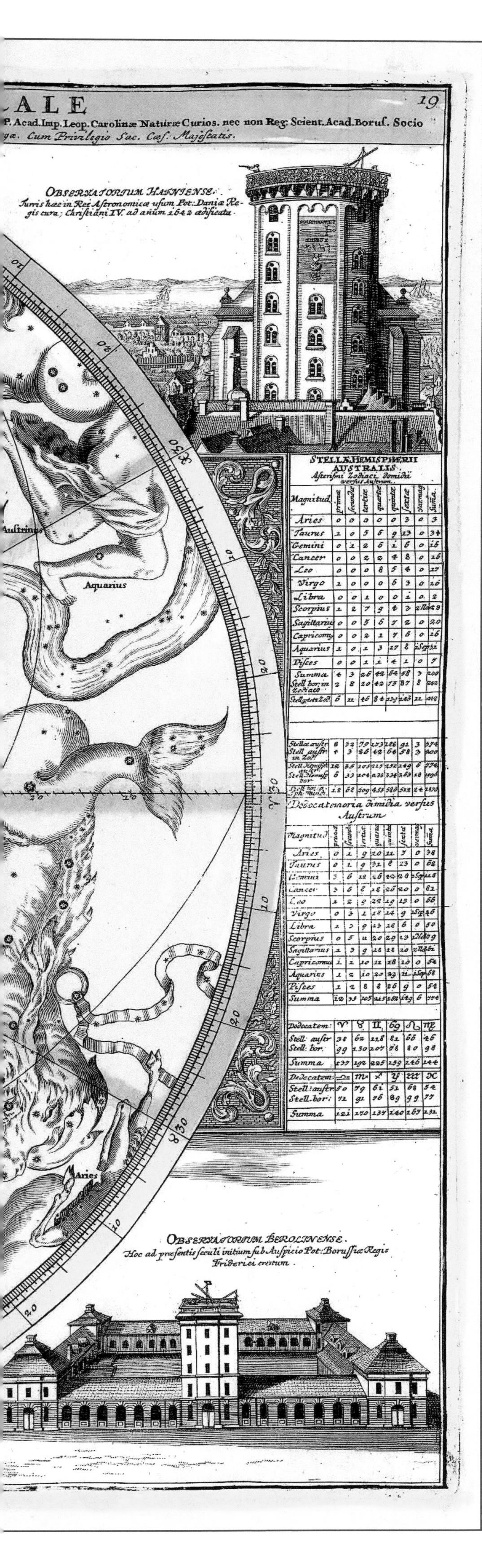

Hemisphaerium Coeli Australe

The map of the southern sky (plate 19 in the atlas) establishes the stars visible in the Southern Hemisphere in 1730. It contains the constellations catalogued by Johannes Hevelius, integrated with the observations of Edmond Halley. Completing the map are two tables with a classification of the stars in the constellations based on their magnitude: at left are the 'austral asterisms', while at right are the constellations of the Zodiac. The plate is embellished with representations in the four corners of the leading European observatories: Greenwich (1676), Copenhagen (1642), Kassel (1714), and Berlin (1711).

Johann Elert Bode

(1747-1826)

Johann Elert Bode was born in Hamburg on 19 January 1747. When still quite young he suffered a disease that seriously damaged his right eye and left him with eyesight problems for the rest of his life. Despite the difficulties created by this handicap, he studied by himself, immediately revealing a gifted mind. This did not escape the notice of his father, who introduced him to the mathematician Johann Georg Büsch, who put his large library at the young man's disposal. Thanks to Büsch's support, when not yet twenty years old Bode was able to publish his first work, a treatise on the eclipse of the Sun that took place on 5 August 1766. Only two years later, in 1768, he published what would be his most successful and popular work: the treatise for amateur astronomers, Anleitung zur Kenntniss des gestirnten Himmels *(*Instructions for Gaining Knowledge of the Starry Heavens*), which went through eight editions up to 1867. In the second edition, which came out in 1772, Bode added the formulation of an empirical rule that indicated the average distances of the planets from the Sun, which in a short time would have a major impact on the history of astronomy, because it made it possible to theorize the presence of still unknown planets. In truth, this principle had already been formulated in 1766 by another scientist, Johann Daniel Titius, but Bode may not have known of its existence and was therefore credited with the discovery. In any case, the law is now universally known as the Titius-Bode Law, thus acknowledging authorship to both scientists.*

The talent revealed by this 25 year-old did not escape notice, and that same year, 1772, the illustrious Swiss physicist Johann Heinrich Lambert invited Bode to teach at the prestigious Berlin Academy of Sciences. In 1774 the two published the first volume of the Astronomisches Jahrbuch *or* Astronomic Yearbook, *which featured the calculations, ephemerides and new discoveries that arrived over the coming years. Despite family tragedies (he married three times, as his first two wives died while giving birth), Bode dedicated himself wholly to astronomic observations all his life. He discovered various galaxies and some comets (even calculating their orbits), but above all he made a decisive contribution to the study of a new planet that was discovered by William Herschel on 13 March 1781 and christened 'Georgium Sidus' (Georgian Star) in honor of King George III of England (Herschel was a German who became a naturalized British citizen). In truth, the celestial body had already been observed by other astronomers who preceded him (for example, John Flamsteed), but all of them, including Herschel, had mistaken it for a star. Whereas Bode declared that it was a planet – the seventh in the solar system – precisely because the criteria of the empirical law concerning the distances between planets that he himself had formulated had been satisfied by his observations of the known planets, from Mercury to Saturn, except for an inexplicable 'gap' between Mars and Jupiter. And in fact the new celestial body bridged that gap perfectly.*

Once it had been established that this was indeed a planet, Herschel proposed naming it Georgian Planet, but the scientific community outside Great Britain objected, sparking a long drawn-out debate. In the end it was Bode who suggested a name taken from mythology: just as Saturn lay 'above' – that is, further away from – Jupiter, and the mythical figure Saturn was the father of this latter, so the new celestial body situated after Saturn should be named after his father Uranus – a suggestion that satisfied everyone.

Besides the Yearbooks, the results of Bode's studies were published in a catalog (1777) and a star map (1782, the original title of which is Vorstellung der Gestirne*) based on the maps published in 1776 by Jean Nicolas Fortin in his version of John Flamsteed's* Atlas Coelestis, *revived and enlarged with the addition of the nebulae and clusters (now known as M92 and M64 in the Messier catalog) he had just discovered.*

His life's work was crowned by his being appointed director of the Berlin Observatory in 1787. Here he could again dedicate his energy to studying Uranus and perfect his knowledge of the cosmos, culminating his oeuvre with the publication, in 1801, of a new star catalog with 17,240 stars, and above all the Uranographia, *a star atlas that combines impeccable scientific accuracy with the artistic beauty of the constellation figures. By now overwhelmed by his eye problems, Bode was forced to leave the Observatory in 1825. On November of the following year he died at the age of 79.*

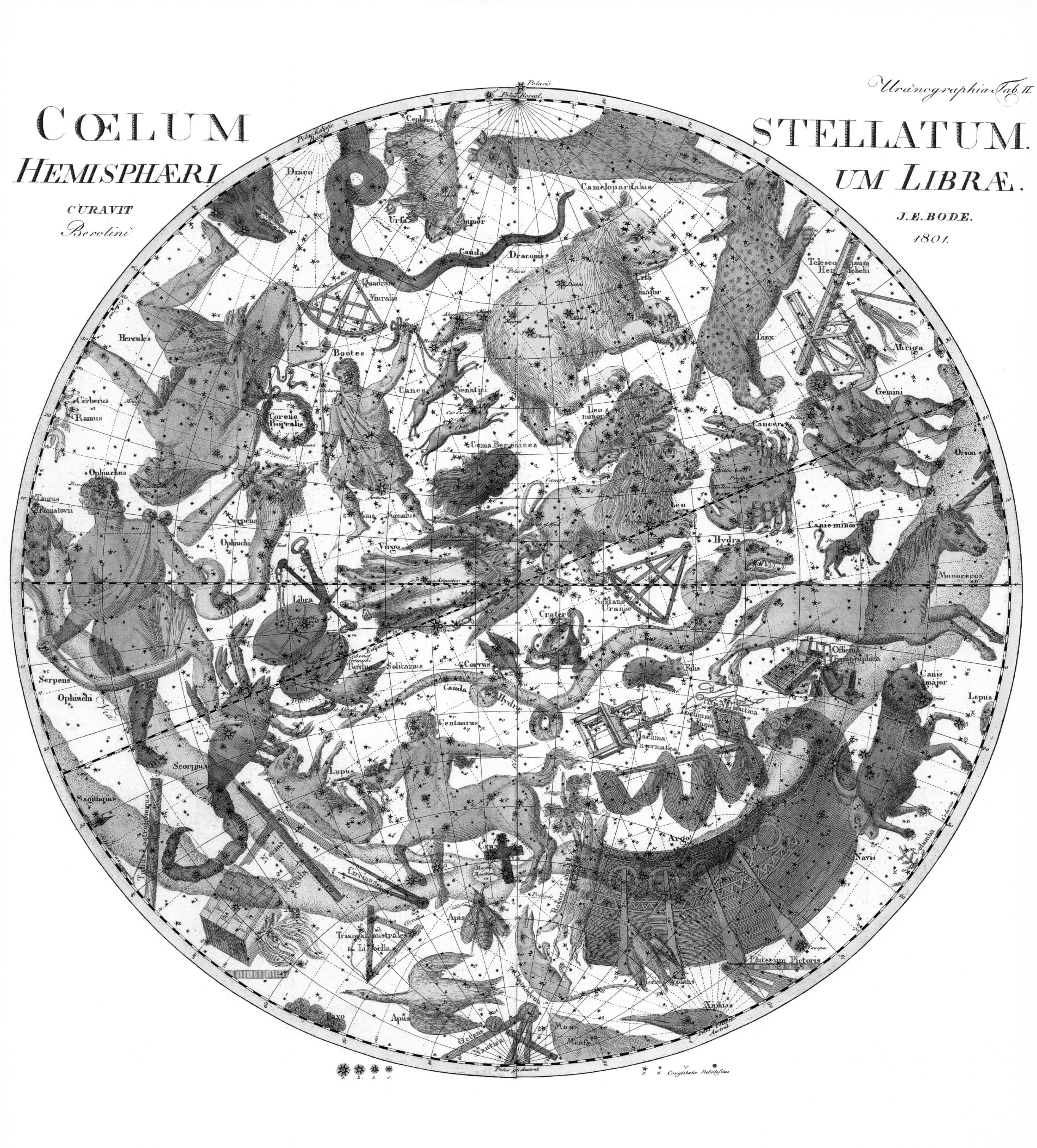
Uranographia Tab. II
CŒLUM STELLATUM.
HEMISPHÆRIUM LIBRÆ.
CURAVIT
Berolini
J.E. BODE.
1801.
Polaris
Polus Boreal.
Cepheus
Draco
Camelopardalus
Ursa minor
Cauda Draconis
Ursa major
Quadrans Muralis
Hercules
Bootes
Canes Venatici
Cerberus et Ramus
Corona Borealis
Leo minor
Cancer
Gemini
Lynx
Auriga
Telescopium Herschelii
Coma Berenicea
Orion
Ophiuchus
Taurus Poniatovii
Serpens
Ophiuchi
Virgo
Leo
Hydra
Canis minor
Monoceros
Libra
Sextans Uraniae
Crater
Corvus
Solitarius
Felis
Officina Typographica
Canis major
Lepus
Serpens
Ophiuchi
Cauda Hydrae
Centaurus
Machina Pneumatica
Scorpius
Lupus
Sagittarius
Argo
Navis
Columba
Crux
Norma
Circinus
Apis
Triangulum australe
Libella
Pictorium Pictoris
Pavo
Apus
Octans
Mons Mensae
Xiphias
Polus Austral.
1. 2. 3. 4.
5. 6. Conglobatio. Nebulositas

Uranographia
(1801)

Johann Elert Bode's *Uranographia sive Astrorum Descriptio* was published in Berlin in 1801 and comprises twenty large format plates with 17,240 stars divided into eight magnitudes, the star clusters known at that time, and 2500 nebulae. More than 100 constellations are also featured, including the decidedly curious ones that were short-lived. Among these latter were *Globus aerostaticus* (aerostat), proposed by Jérôme Lefrançois de Lalande in 1798 as a tribute to the Montgolfier brothers, pioneers in hot-air ballooning; *Machina electrica*, representing an electrostatic generator; *Lochium funis* (the Log and Line, two navigation instruments); *Officina typographica*, which commemorated the 350th anniversary of Johannes Gutenberg's invention of the movable type press; and *Honores Friderici* (Glory of Frederick), in honor of King Frederick II of Prussia, who died in 1786, a few years before the atlas was published. Bode was also the first person to use dotted lines to mark the outline of the constellations an innovation that was modified and perfected in the first decades of the 20th century.

His *Uranographia* is at once the pinnacle and the 'swansong' of classical celestial cartography. In fact it was the last atlas conceived with the aim of combining the most accurate and complete scientific data possible with the purely aesthetic need to depict the stars artistically. For that matter, the growing number of new stars discovered due to the incessant progress in technology made it more and more difficult to 'draw' the constellations in keeping with traditional iconography. Thus, by the mid-18th century the mythological figures would become progressively stylized, until they ended up disappearing completely in scientific atlases, being replaced by purely technical representations. Traditional illustrations would continue to be used for a few decades more in non-specialist atlases, but with a merely decorative function.

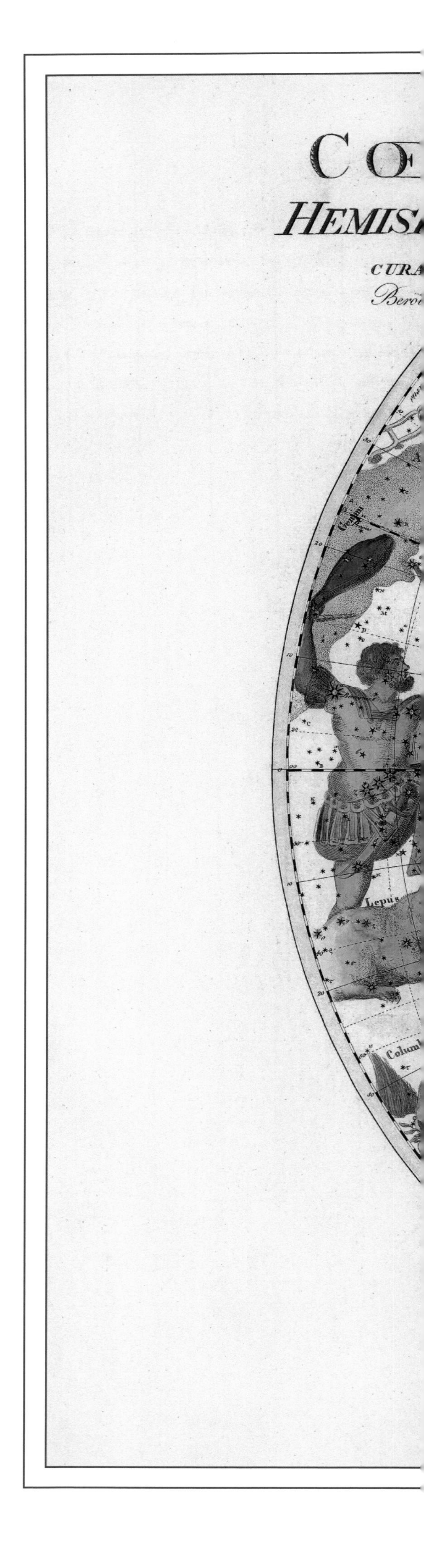

Coelum Stellatum

As was the custom in all atlases, Bode proposed the usual two celestial hemisphere maps, but with one basic difference. Instead of focusing on the representation of the Northern and Southern Hemisphere skies, he chose to use a 'frontal' view that concentrates on the Libra (Scales) and Aries (Ram) constellations. Because of the continuous astronomical discoveries and the new constellations introduced by Lalande and Bode himself, by the year 1801 both hemispheres (the other is on the preceding page) had become quite crowded.

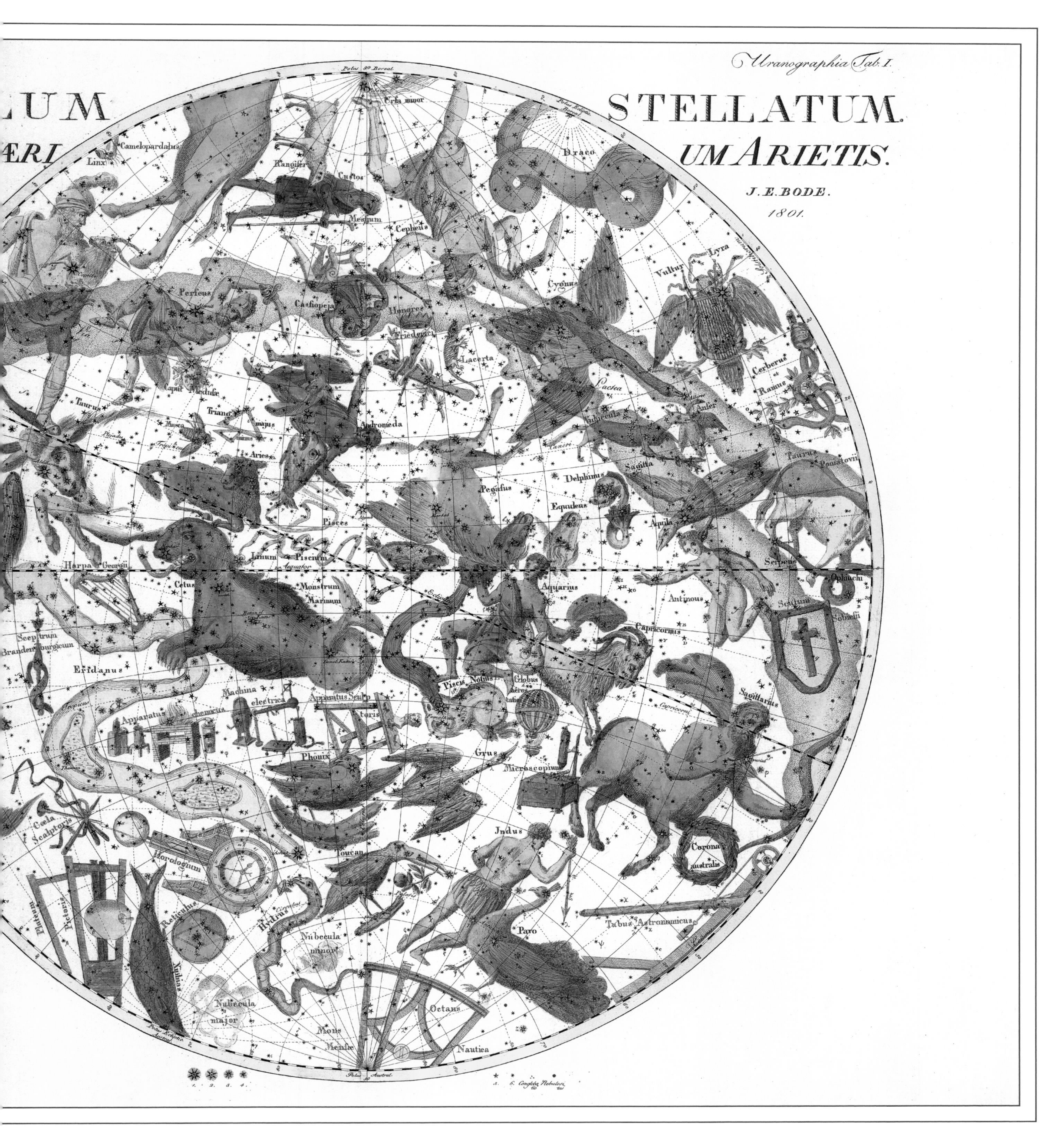

Uranographia Tab. I.
LUM
STELLATUM.
ÆRI
UM ARIETIS.
J. E. BODE.
1801.
Polus 90 Boreal
Ursa minor
Camelopardalus
Linx
Rangifer
Custos
Messium
Cepheus
Draco
Lyra
Vultur
Cygnus
Perseus
Cassiopeja
Honores
Friederici
Lacerta
Cerberus
Ramus
Lactea
Anser
Capit Medusæ
Taurus
Triang.
Musca
Aries
Andromeda
Vulpecula
Sagitta
Delphinus
Pegasus
Equuleus
Aquila
Taurus Poniatovii
Pisces
Linum Piscium
Aequator
Harpa Georgii
Cetus
Monstrum Marinum
Aquarius
Antinous
Serpens
Ophiuchi
Scutum Sobieski
Capricornus
Sceptrum Brandenburgicum
Eridanus
Machina electrica
Apparatus Sculptoris
Apparatus Chemicus
Piscis Notius
Globus aerostaticus
Sagittarius
Capricorni
Phœnix
Grus
Microscopium
Cæla Scalptoris
Horologium
Toucan
Indus
Corona australis
Reticulus
Hydrus
Pictoris
Nubecula minor
Tubus Astronomicus
Pavo
Xiphias
Nubecula major
Mons Mensæ
Octans
Nautica
Polus Austral.
1. 2. 3. 4. 5. 6. Conglob. Nebulosi

Honores Friderici, Andromeda, etc.

In the 1787 issue of Astronomisches Jahrbuch *Bode introduced a new constellation that he had conceived and dedicated to Frederick II the Great, King of Prussia, who had died the preceding year. Bode called it* Friedrichs Ehre, *the honor and glory of Frederick (*Honores *or* Gloria Friderici *in Latin), and depicted it as a sword with a crown, accompanied by a goose quill and an olive branch that symbolize the wisdom and justice of the monarch. This constellation, situated between the right arm of Andromeda and the Lacerta (Lizard) constellation (introduced by Hevelius in 1687), was used again by Bode in* Uranographia, *replacing Sceptrum et Manus Iustitiae (Scepter and Hand of Justice), which was created in 1679 by the Frenchman Augustin Royer to honor King Louis XIV. Both constellations are now obsolete.*

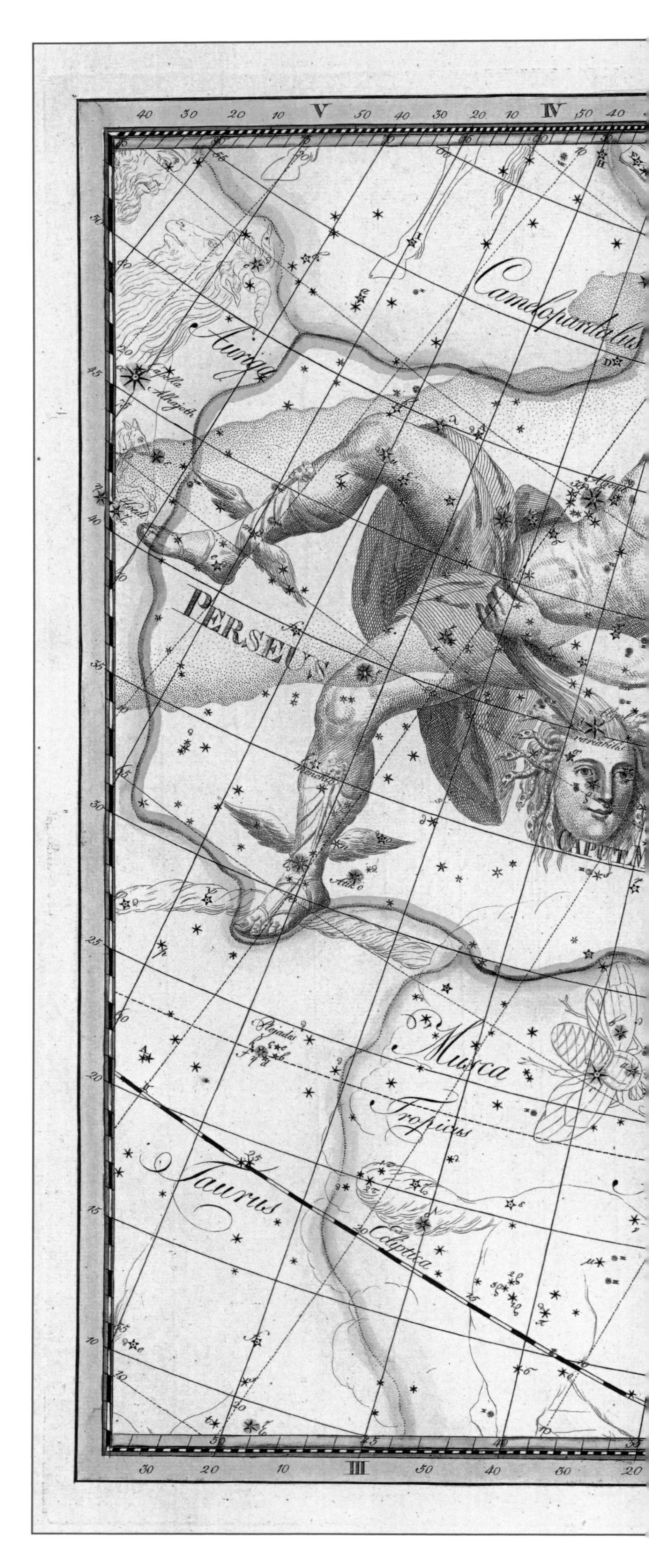

IV.

III II I XXIV XXIII XXII XXI

CASSIOPEJA

Cepheus

Cygnus

Lactea

Colurus

Caph

Schedir

LACERTA

HONORES FRIDERICI

Adhil

ANDROMEDA

Mirach

Pegasus

Scheat

TRIANGULUM MAJUS

MINUS

Cancri

Markab

Aequinoctior.

Algenib

Pisces

Mesarthim

II I XXIV

Draco, Custos Messium, Rangifer etc.

In this section of the northern sky Bode reproduced the constellations Draco (Dragon-headed Serpent), Cepheus (Soldier), Ursa Minor (Small Bear), Camelopardalis (Giraffe), Rangifer (Reindeer) and Custos Messium (Harvest Keeper). The last-mentioned constellation was introduced by Joseph Jérôme de Lalande in 1775 as a tribute to his colleague Charles Messier, who discovered and designated numerous comets. In 1801 the British scientist Thomas Young renamed the constellation Vineyard Keeper. However, Bode preferred to retain the French astronomer's original name, placing it next to Rangifer, which had been introduced in 1743 by Le Monnier as Le Renne but was given its Latin name by Bode. Both are now obsolete, while Draco is not. This latter, already seen by Ptolemy, represents the dragon Ladon that in Greek mythology guarded the Hesperides garden and was killed by Hercules.

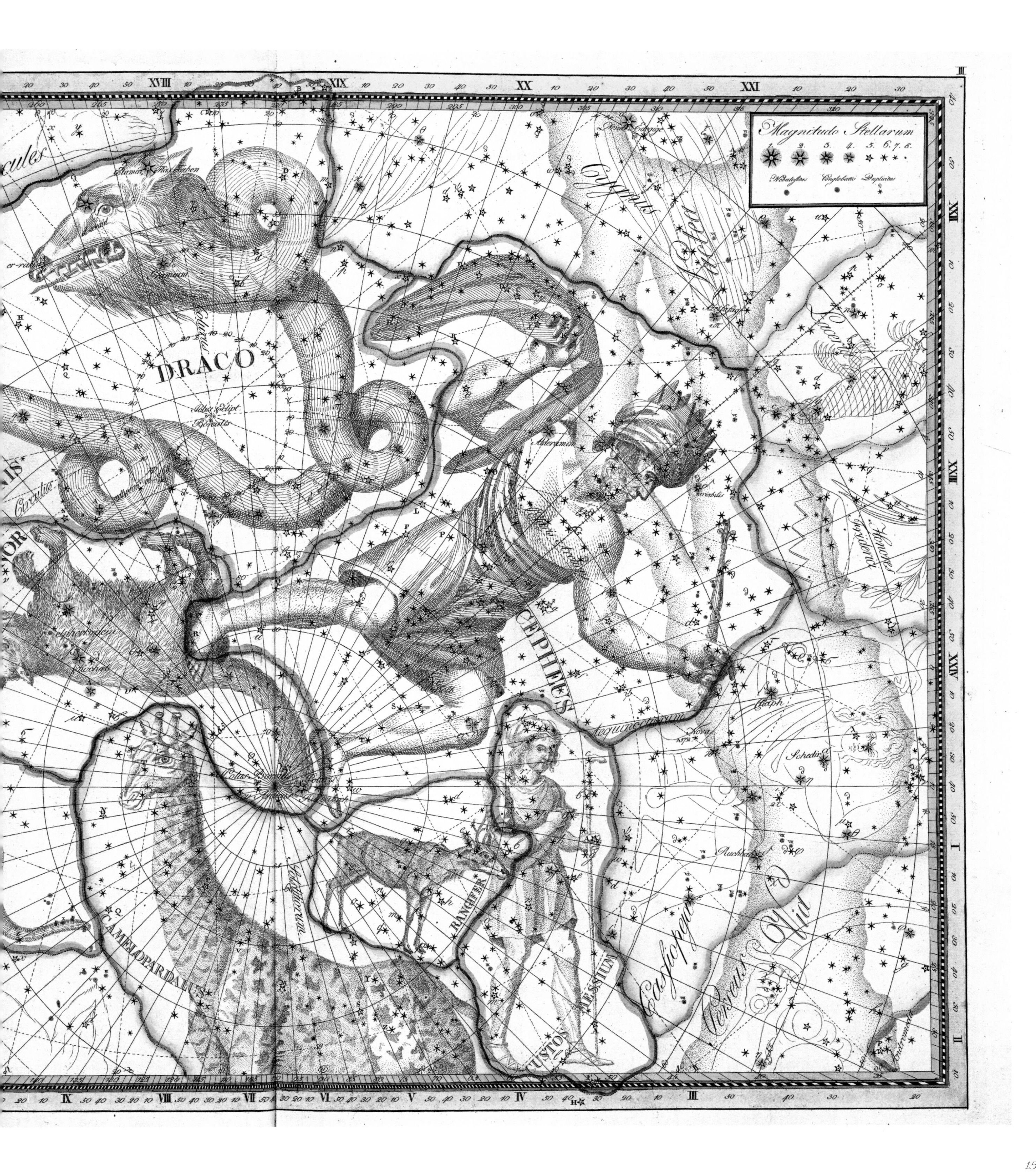

Magnitudo Stellarum
DRACO
CEPHEUS
CAMELOPARDALUS
RANGIFER
CUSTOS MESSIUM
Cygnus
Lacerta
Cassiopea
Perseus
Alderamin
Schedir
Ruchbah

Harpa Georgii, Gemini, Orion, Taurus, etc.

Besides the well-known zodiacal constellations Gemini (Twins), Taurus (Bull) and Orion, this chart in the Uranographia *also has Harpa Georgii (George's Harp), introduced only a few years earlier (1789) by the Hungarian astronomer Maximilian Hell, director of the Vienna Observatory, to honor King George III of England, the patron of William Herschel, who discovered the planet Uranus. Hell had originally called it Psalterium Georgianum (George's Psaltery), but Bode changed it to Harpa Georgii, discarding the ancient psaltery – a stringed instrument no longer used in Bode's time – and replacing it with the much more versatile harp (the pedals of which had been greatly improved), which precisely in this period became extremely popular. Harpa Georgii is now obsolete.*

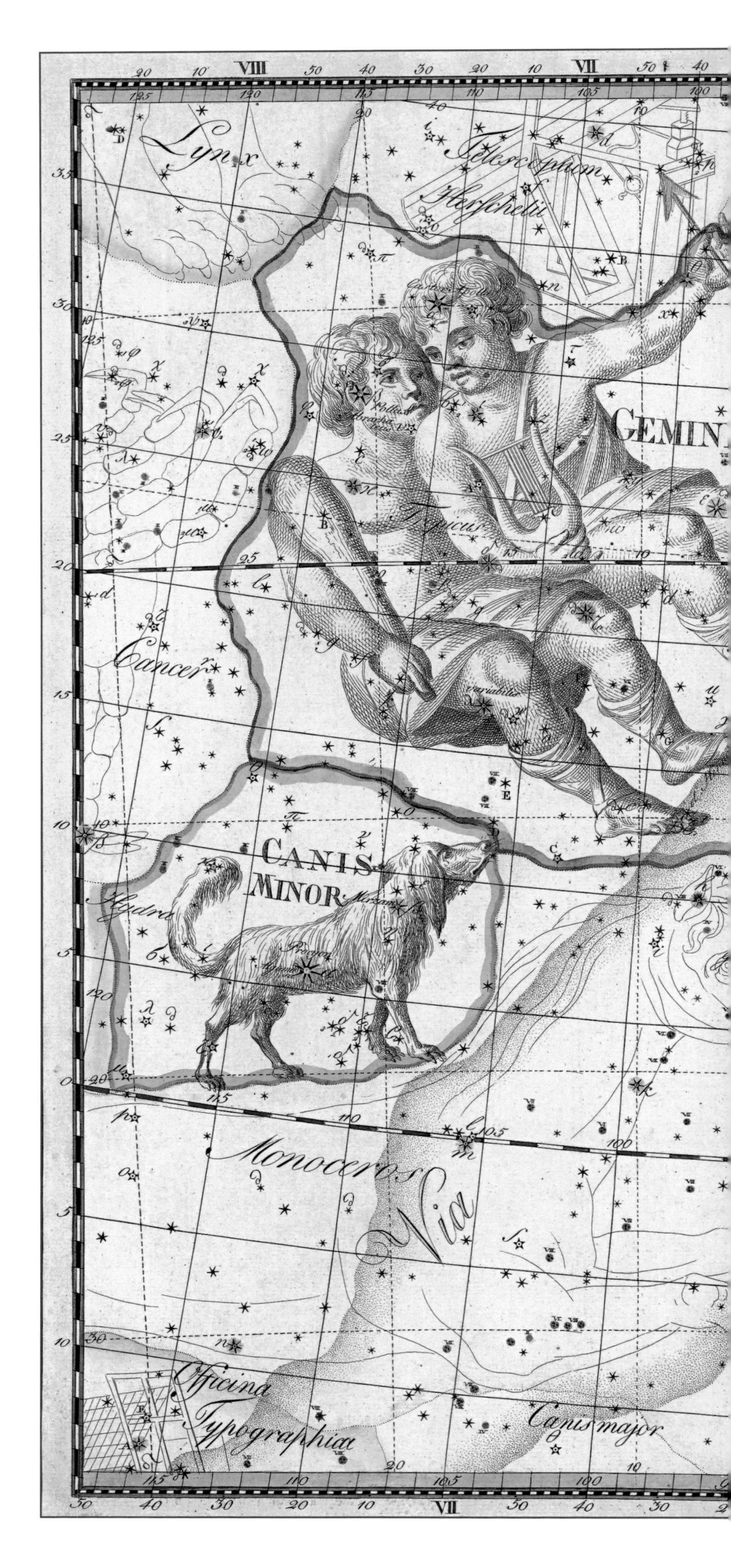

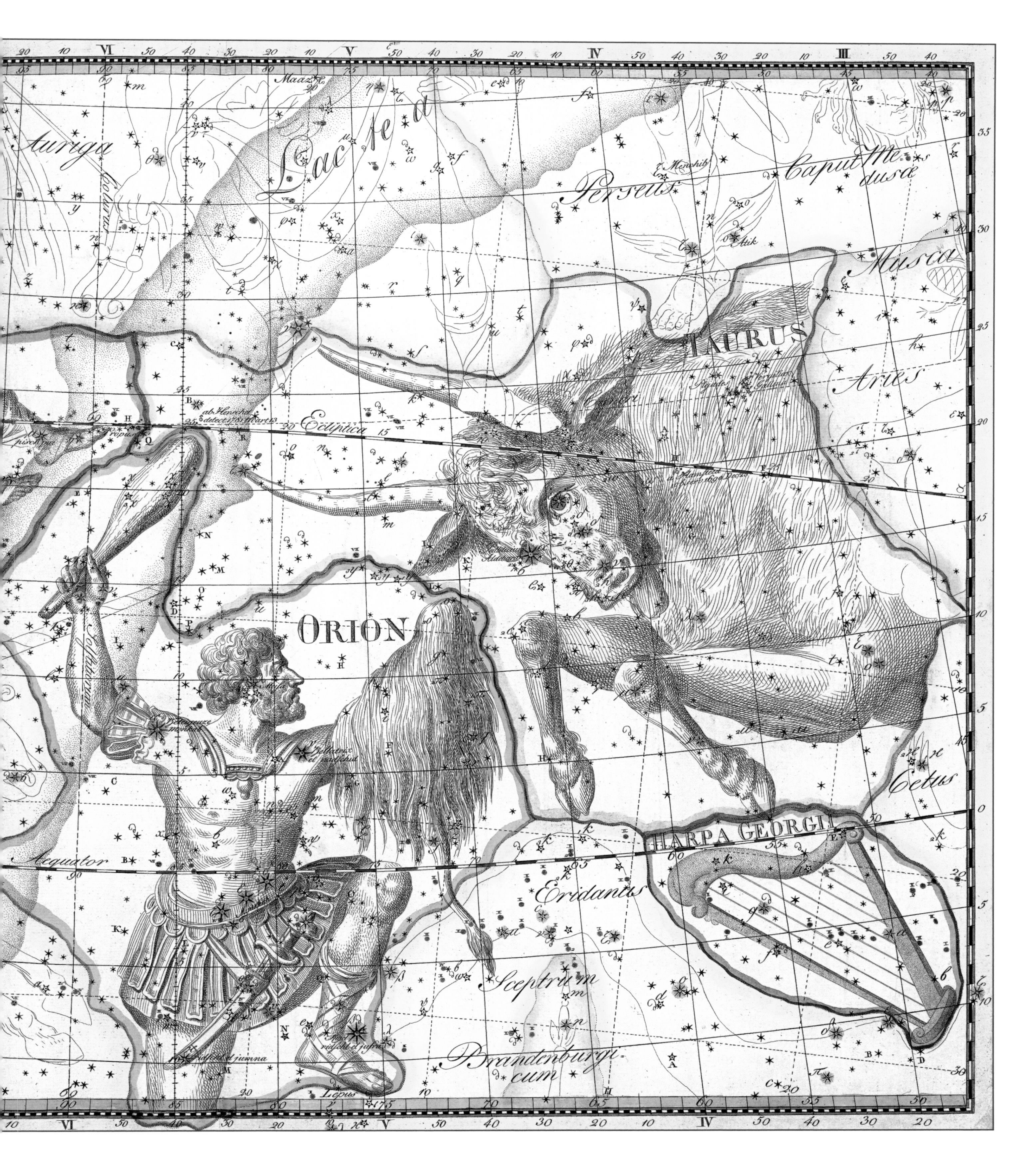

Auriga
Lactea
Perseus
Caput Medusæ
Musca
TAURUS
Aries
Ecliptica
ORION
Cetus
HARPA GEORGII
Eridanus
Sceptrum
Brandenburgicum
Aequator
Lepus

Turdus Solitarius, Virgo, Libra

*In this map Bode depicts the well-known constellations Virgo (Virgin) and Libra (Scales), as well as the newborn Turdus Solitarius (Solitary Thrush). This latter was introduced in 1776 by Pierre Charles Le Monnier to occupy the 'empty' area near the tail of Hydra, resembling an exotic bird "from the Indies and Philippines," probably the Rodrigues solitaire (*Pezophaps solitaria*). Widespread in the Mascarene Island (Indian Ocean), it was similar to the dodo and like this latter was flightless (the species is extinct). Bode chose to depict Turdus as a blue rock thrush (*Monticola solitarius*), which was widespread in Europe. Although some astronomers proposed other names for the constellation, today it is considered obsolete and is no longer in the star charts.*

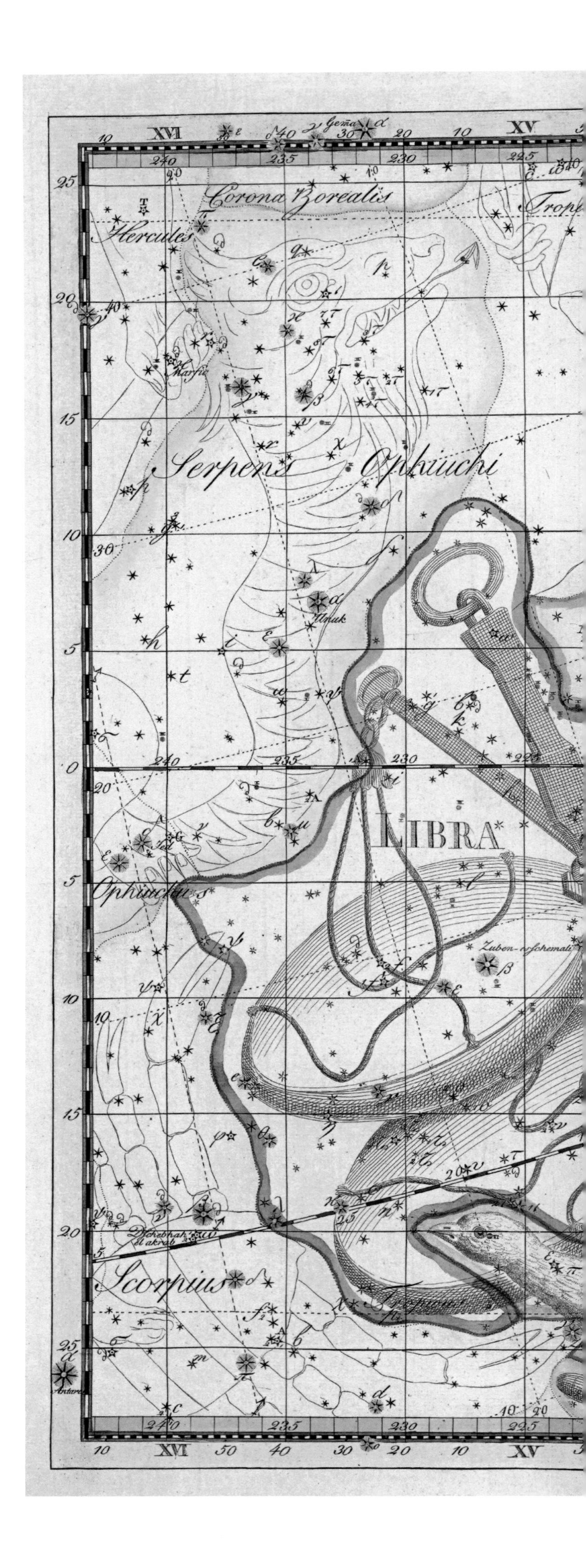

Coma Berenices
Leo
Bootes
Mons Mænalus
VIRGO
Crater
Corvus
TURDUS SOLITARIUS
CAUDA HYDRÆ seu Serpentis aquatici
Centaurus
Aequator
Ecliptica
Arcturus
Vindemiatrix
Spica

Alexander Jamieson

(1782-1850)

Alexander Jamieson was born in 1782 in Rothesay, Scotland. Very little is known about his childhood except that he was the son of a craftsman. It is almost certain that his studies were irregular; in fact, he earned a degree at Aberdeen when he was almost forty and in 1825 enrolled at St. John's College, Cambridge to study theology, but whether or not he graduated is still obscure. In any case, in the meantime he became a schoolmaster in numerous institutes: first at Kensington until 1822, then at the Heston House Classical and Mathematical School near Hounslow, west of London, and lastly at the Wyke House Academy in Middlesex, where he stayed until it went bankrupt in 1838.

Concomitant with his teaching activity Jamieson produced many scholastic textbooks, including two important grammars – A Grammar of Rhetoric and Polite Literature *(1818) and* A Grammar of Logic and Intellectual Philosophy *(1819) – that were extremely successful and went through many editions and reprints, especially in the United States. Before these works, in 1814 the Darton, Harvey and Co. publishing house issued his* A Treatise on the Construction of Maps, *which was also fairly popular.*

In March 1820 Jamieson married Frances Thurtle, who also wrote didactic books, which were published by G. & W.B. Whittaker of London. Two years later the same publishing house printed Jamieson's second cartographic work, A Celestial Atlas, Comprising a Systematic Display of the Heavens…, *which was a great success and went through many editions, becoming famous; furthermore, thanks to this work, in 1826 he was made a member of the London Astronomical Society.*

Certain that he could exploit his popular atlas to the full, even for educational purposes, Jamieson set out to produce a version aimed at schools. The result was a celestial atlas – titled An Atlas of Outline Maps of the Heavens *(1824) – with only the outlines of the mythological constellations, and no indications of the stars and a graticule. The idea was that after having studied the original and complete* Celestial Atlas, *the students would have to position the stars on the plate by making use of the graticule, conforming to both the coordinates and the magnitude.*

Despite the fact that his courses were renowned – advertisements that announced them appeared in some of the leading newspapers of the time – in March 1838 the Wyke House School went bankrupt and Jamieson found himself without means. He was therefore obliged to seek another job, becoming an actuary, analyzing statistics to calculate risks, reserves and premiums for life insurance firms. Despite the fact that this was certainly not his métier, he managed to take advantage of this experience to write, in 1841, a book on the constitution and operations of life insurance companies, which was his last work. By now a forgotten figure who had become almost poverty stricken, in 1849 Jamieson suffered a stroke, which undermined his health. He moved to Bruges, Belgium with his wife in the hope of being able to change the course of his life, and died there on 6 July 1850 at the age of 68.

Stereographic Projection of the Northern Celestial Hemisphere on the Plane of the Equinoctial

This celestial map illustrates all the visible constellations in the northern sky. Each is accompanied by its magnitude according to the scale in the lower right legend.

Published Feby 1. 1822 by G & W. B. Whittaker T. Cadell, & H. Hailes, London.

A Celestial Atlas, Comprising a Systematic Display of the Heavens (1822)

Alexander Jamieson's *Celestial Atlas* was published in London in February 1822 by G & W.B. Whittaker. This work, dedicated to King George IV of England, clearly seems to have drawn inspiration from the *Atlas Coelestis* published in 1729 by John Flamsteed, from Johann Elert Bode's *Vorstellung der Gestirne* (1782) and above all from his *Uranographia* (1801). In particular, the similarities between Jamieson's and Bode's atlases are clearly evident. Indeed, for his plates the Scotsman utilized the same size chosen by the German astronomer, 9 x 7 inches (22.5 x 17.5 cm), and the number of plates is identical – thirty. On the other hand, the most striking difference lies in the representations of the constellations, which in Jamieson's atlas are rendered much more meticulously: the superior aesthetic quality and greater appeal is especially appreciable in the rendering of the figures, which in Flamsteed's edition seem crude and poorly executed, for example in the Canis Major.

The *Celestial Atlas* enjoyed an immediate and great success, so much so that it sold out in a short time, and a second edition came out only seven months later, wholly identical to the first (even including the misprints) except for the addition of another plate, inserted after the Preface, that illustrates the Hindu and ancient Egyptian Zodiacs.

Of the thirty plates, twenty-six are star charts, two represent the celestial hemispheres, one features the brightest stars in the Northern Hemisphere, while the last one illustrates the Moon, Venus, Jupiter and Saturn. They were originally supposed to be larger, but the author chose to reduce the size to limit the printing costs. In some versions of the atlas the plates are hand-colored.

As for the constellations, the plates contain more than one hundred, including three that Jamieson himself 'invented': Norma Nilotica, with the figure of Aquarius holding a rod to measure the depth of the Nile River; Noctua the night owl; and, in the southern sky, Solarium (Sundial), which replaced Reticulum (Net), introduced by Nicolas-Louis de Lacaille in 1763.

Boötes, Canes Venatici, Coma Berenices, Quadrans Muralis

While at work on his atlas, Jamieson personally saw to the artistic aspect of the charts, as he was determined to make the figures in the constellations aesthetically more elegant than those produced by Bode. The copper plates were executed by the London engravers Neele and Son, who were among the best on the market at this time. And in fact, compared to Bode's atlas, as fine as it was, Jamieson's figures (such as Boötes and the Hunting Dogs, depicted here) are much more elegant and expressive.

Mural Quadrant
BOOTES
CANES VENATICI
COMA BERENICE
Asterion
Chara
Cor Caroli
Benetnasch
Alcor
Mizar
Alioth
Merga
Mirac
Arcturus
Alkalurops
Mons Mænalus
Vindemiatrix
Ursa
COLURE
Æ in Deg.
Æ in Time

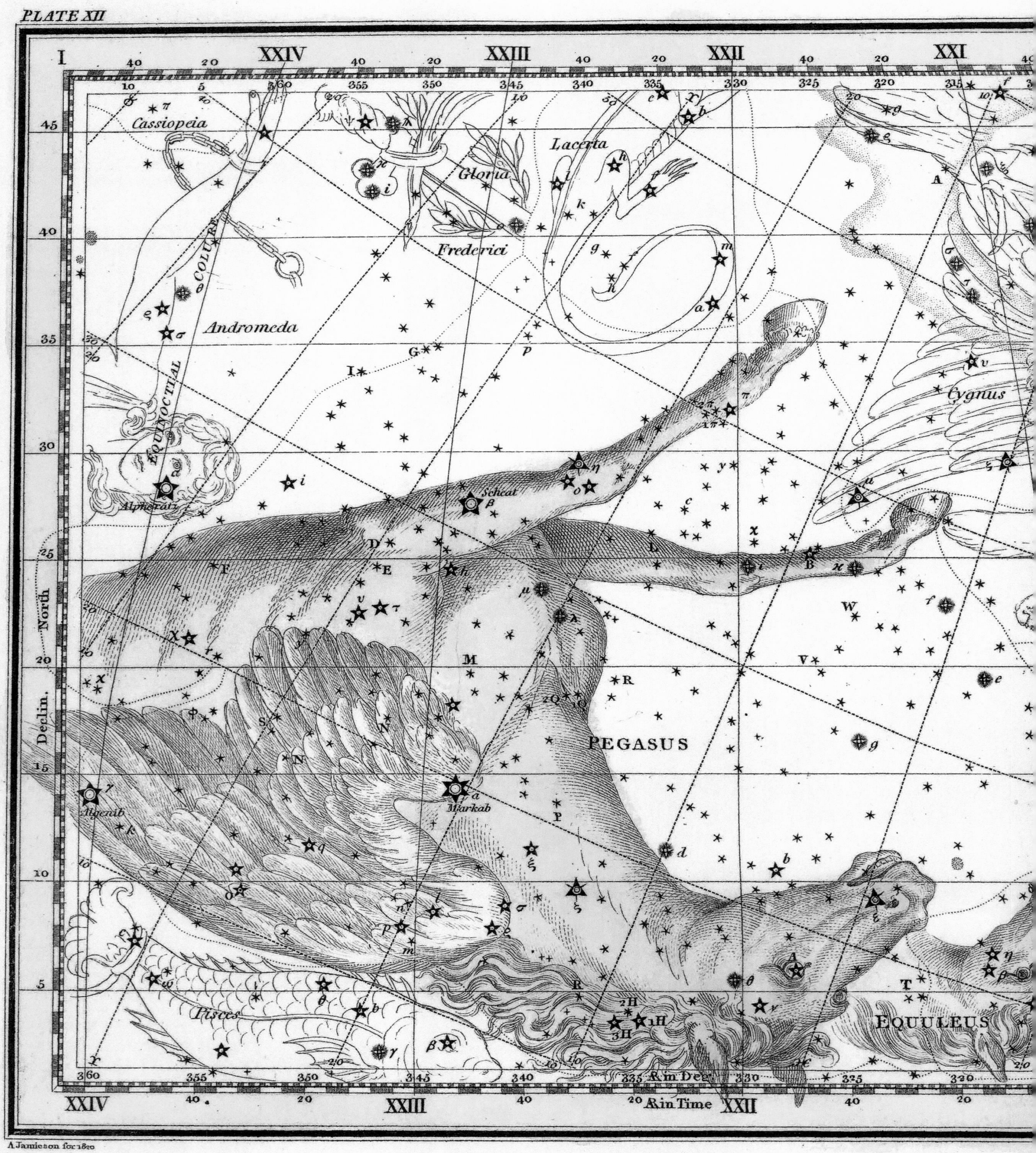

PLATE XII

Cassiopeia

Lacerta

Gloria Frederici

Andromeda

Cygnus

EQUINOCTIAL COLURE

Alpheratz

Scheat

PEGASUS

Algenib

Markab

Pisces

EQUULEUS

North Declin.

Æ in Deg.

Æ in Time

A. Jamieson fec. 1820

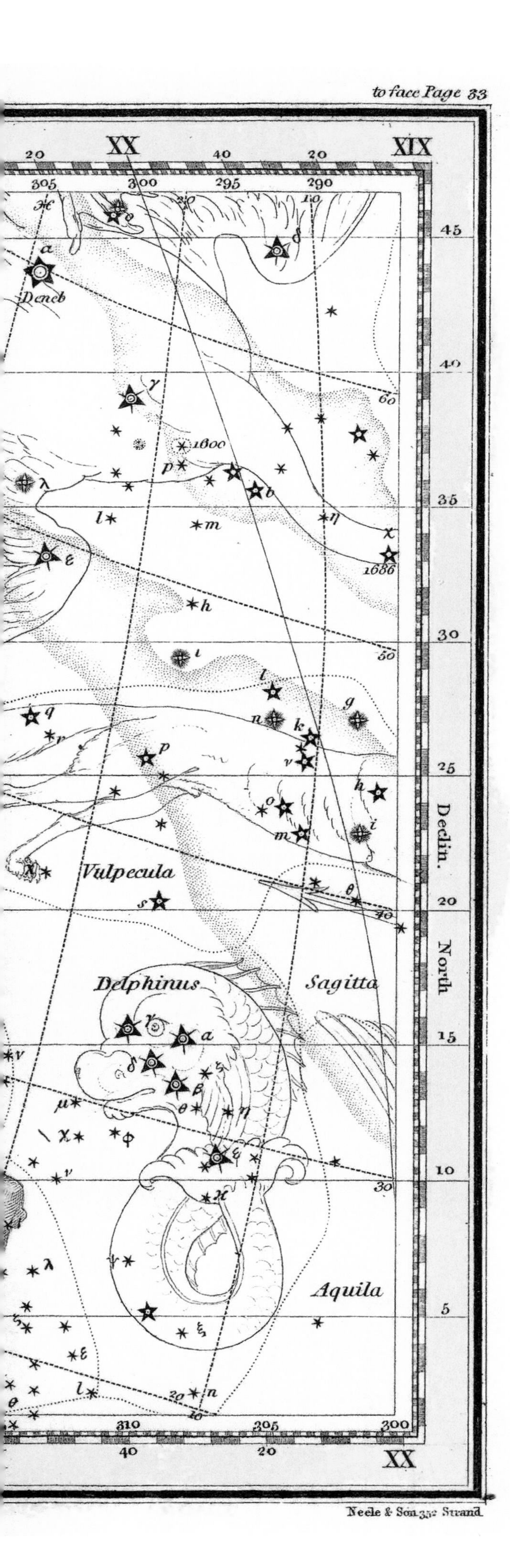

Pegasus

This plate is for the most part dedicated to Pegasus, the famous Northern Hemisphere constellation. Standing out at left are the four stars that comprise the asterism known as the Square of Pegasus. The stars are Markab (Alpha Pegasi), Scheat (Beta Pegasi), Algenib (Gamma Pegasi) and Sirrah, which in the past was considered both part of Pegasus and of Andromeda, but is now attributed only to the latter and in fact is called Alpha Andromedae. All these stars have Arabic names: Sirrah or 'horse's navel', Algenib or 'flank', Scheat or 'shin' (but sometimes also known as Sheat Alpheras or 'horse's shin'), Markab or 'saddle'. They are all represented in Jamieson's plate.

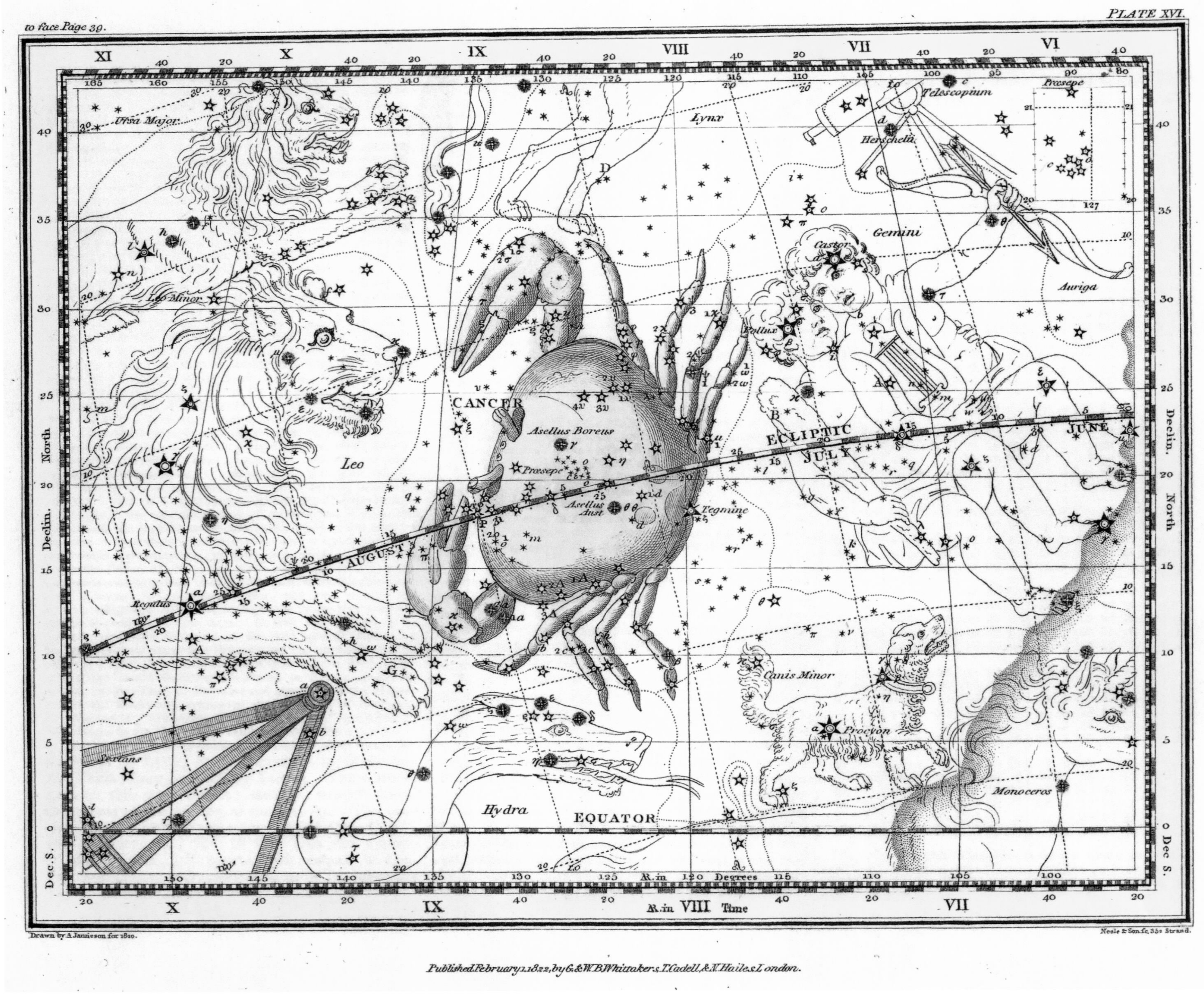

Cancer

From an aesthetic standpoint, the zodiacal constellation Cancer or Crab, like all the others that Jamieson depicted, is well executed and rich in details. The large crab is modeled masterfully and is vibrant, 'alive' and three-dimensional thanks to the skillful play of chiaroscuro, which is even more apparent in the versions with hand-colored plates. The figures of the constellations are separated by a light dotted line that marks out their 'boundaries'. Although Jamieson depicted a much larger number of stars than all the preceding celestial atlases, the plates of the constellations are by no means crammed; on the contrary, they impart an atmosphere of order and lightness.

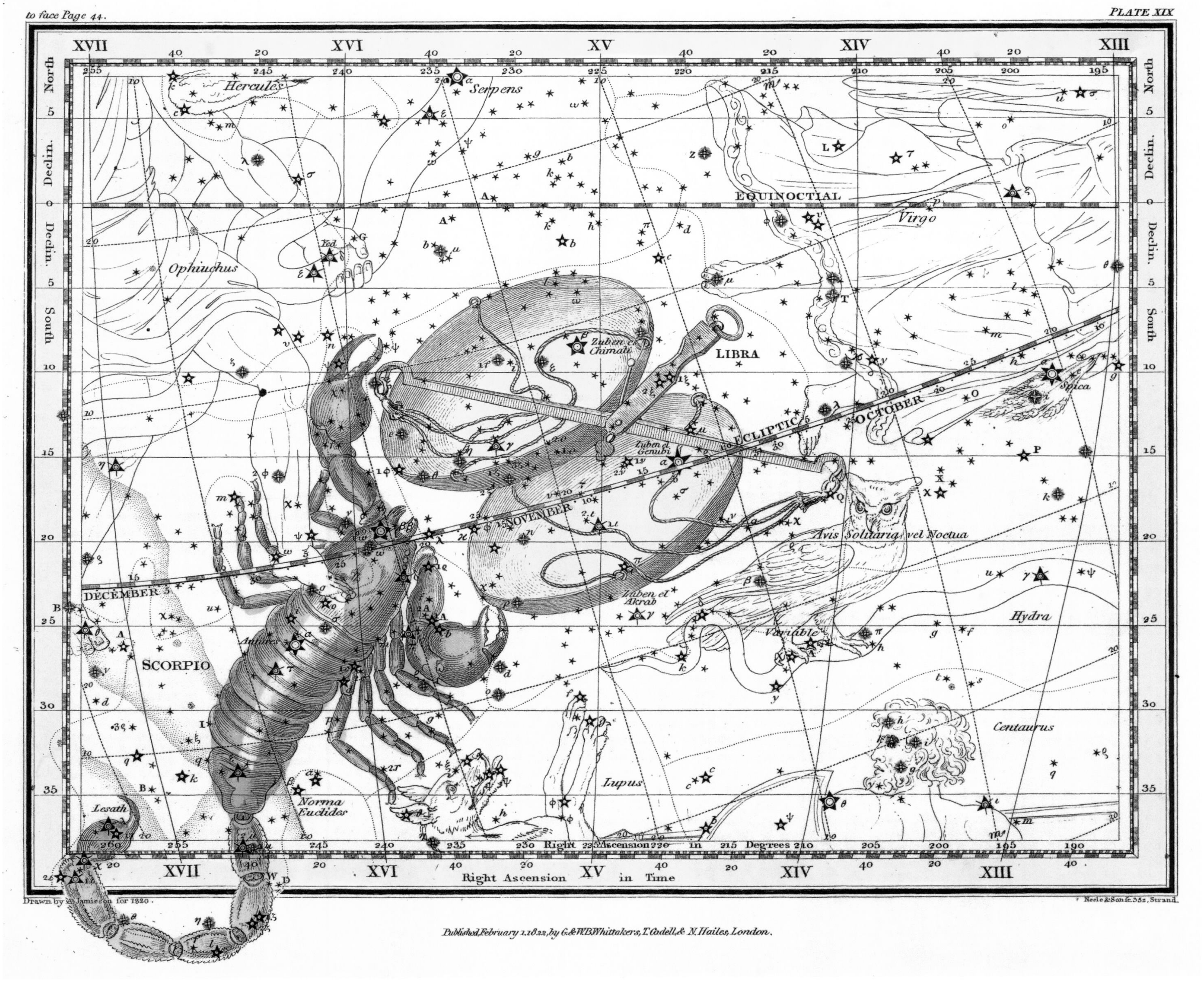

Scorpio et Libra

The aesthetic observations made regarding the preceding plate of Cancer also apply to this one, which highlights two other zodiacal constellations, Scorpion and Balance. Precisely because of its didactic character, Jamieson's atlas does not neglect listing in detail the names of the best known and most visible stars in the sky. A case in point is Antares (Alpha Scorpii), which stands out in the middle of Scorpio, in correspondence to its heart (Calbalacrab in Arabic). This is a red supergiant star that is 604 light years from our solar system; its radius is about 850 larger than that of the Sun, which makes it one of the largest known stars. Jamieson also illustrates Libra (Balance), which is next to Scorpio; in fact, it was originally a part of this latter, representing its pincers.

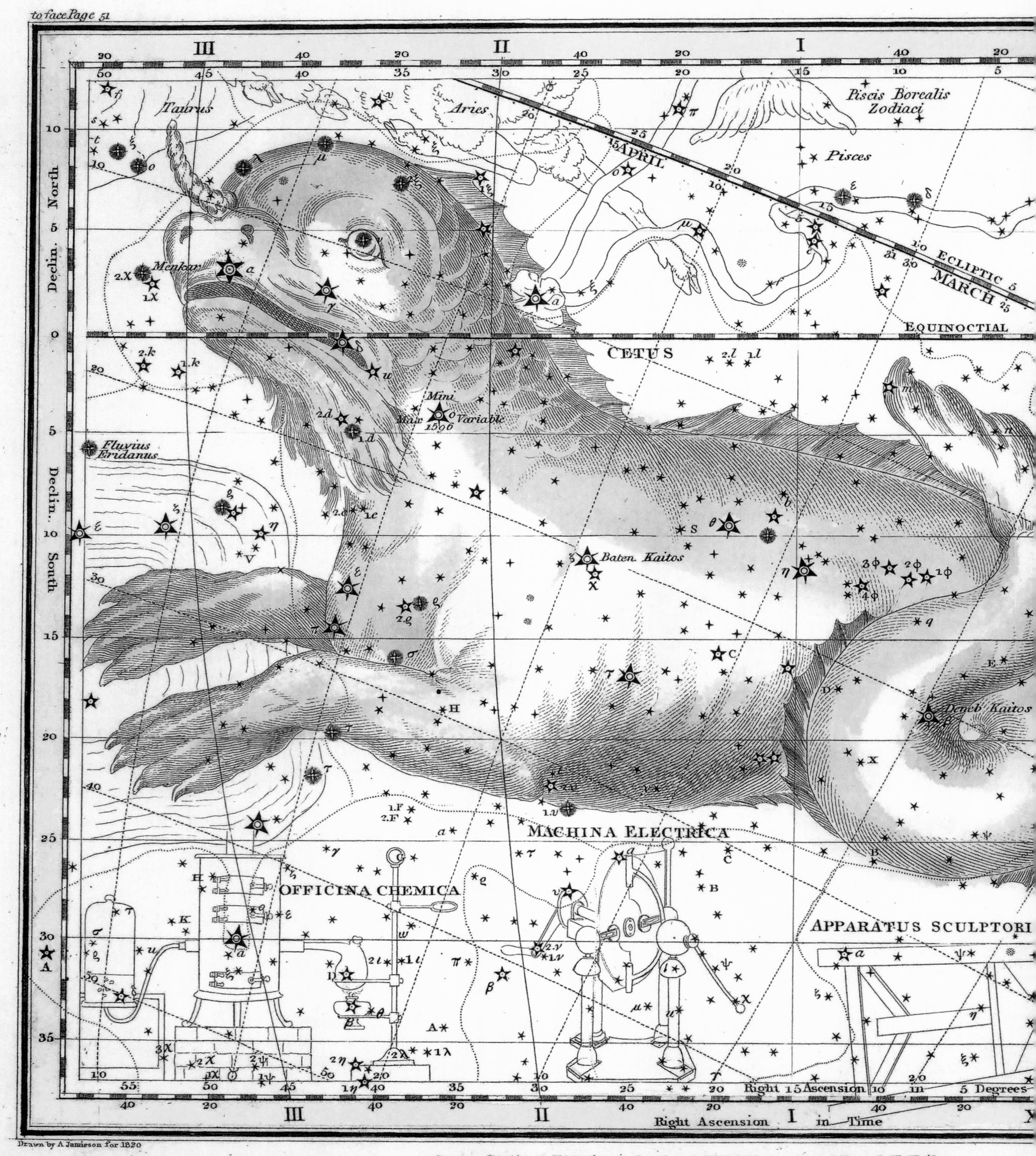

to face Page 51
Taurus
Aries
Piscis Borealis Zodiaci
Pisces
APRIL
MARCH
ECLIPTIC
EQUINOCTIAL
Menkar
CETUS
Mini
Max
Variable
1596
Fluvius Eridanus
Baten Kaitos
Deneb Kaitos
MACHINA ELECTRICA
OFFICINA CHEMICA
APPARATUS SCULPTORI
Declin. North
Declin. South
Right Ascension in Degrees
Right Ascension in Time
Drawn by A. Jamieson for 1820
London, Published Feby. 1st. 1822 by G. and W. B. Whitaker. T. Cadell and N. Hailes.

Cetus

This plate depicts the constellation of Cetus, the sea monster that in Greek mythology Perseus killed in order to free the beautiful Andromeda, who was condemned to be devoured by Cetus. Drawing from the iconography already used by Cellarius, Jamieson decided to depict it differently, not as a dragon (as is appears in Bayer's atlas, for example), but rather as a creature similar to a whale, albeit with front paws and a horn. Below Cetus's body Jamieson chose to represent three of the modern constellations conceived to celebrate recent technological achievements: from right to left, Sculptor, Machina Eletrica (Electric Generator), and Fornax Chimiae (or Chemical Furnace, today known as Fornax, introduced by Nicolas-Louis de Lacaille in 1756).

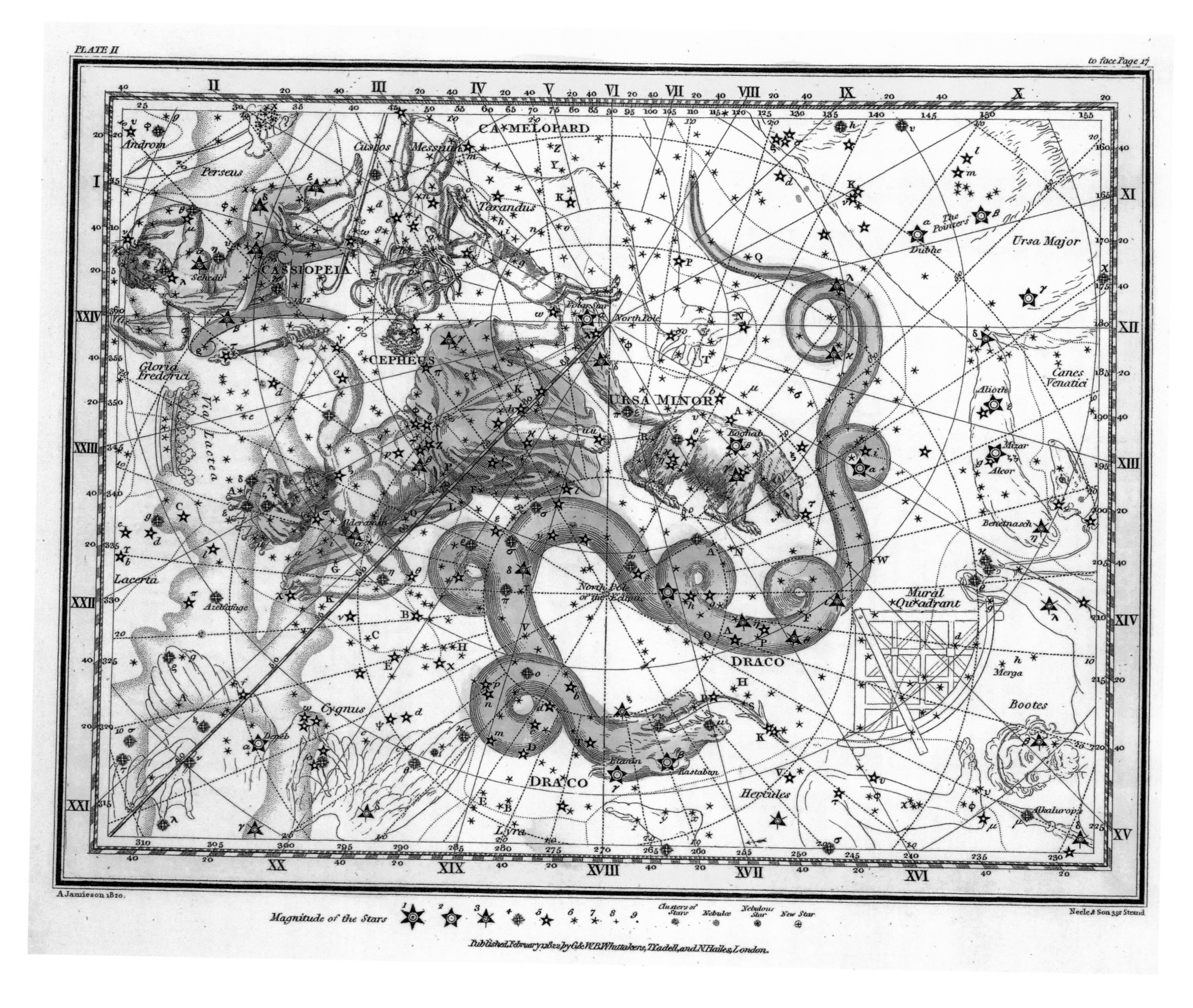

Draco, Ursa Minor, Cepheus, Cassiopeia

Among the best known constellations in the Northern Hemisphere, partly because it is easy to identify thanks to its zigzag shape, is Cassiopeia, named after the vain wife of the King of Ethiopia Cepheus (who appears next to her). One day she boasted she was more beautiful than the Nereid sea nymphs, irritating them so much that they asked Poseidon to intervene. Thus the god of the sea sent the sea monster Cetus to plague the coasts of the Ethiopian kingdom until it would be placated by the sacrifice of the Princess Andromeda to Cetus; as was mentioned above, she was saved by Perseus. Cassiopea, depicted here admiring herself in a mirror, was condemned to wander forever in the sky upside down. Note Jamieson's interpretation of the Gloria Frederici constellation (now obsolete) in this plate: he eliminates the sword and other attributes, reducing the representation to the crown alone.

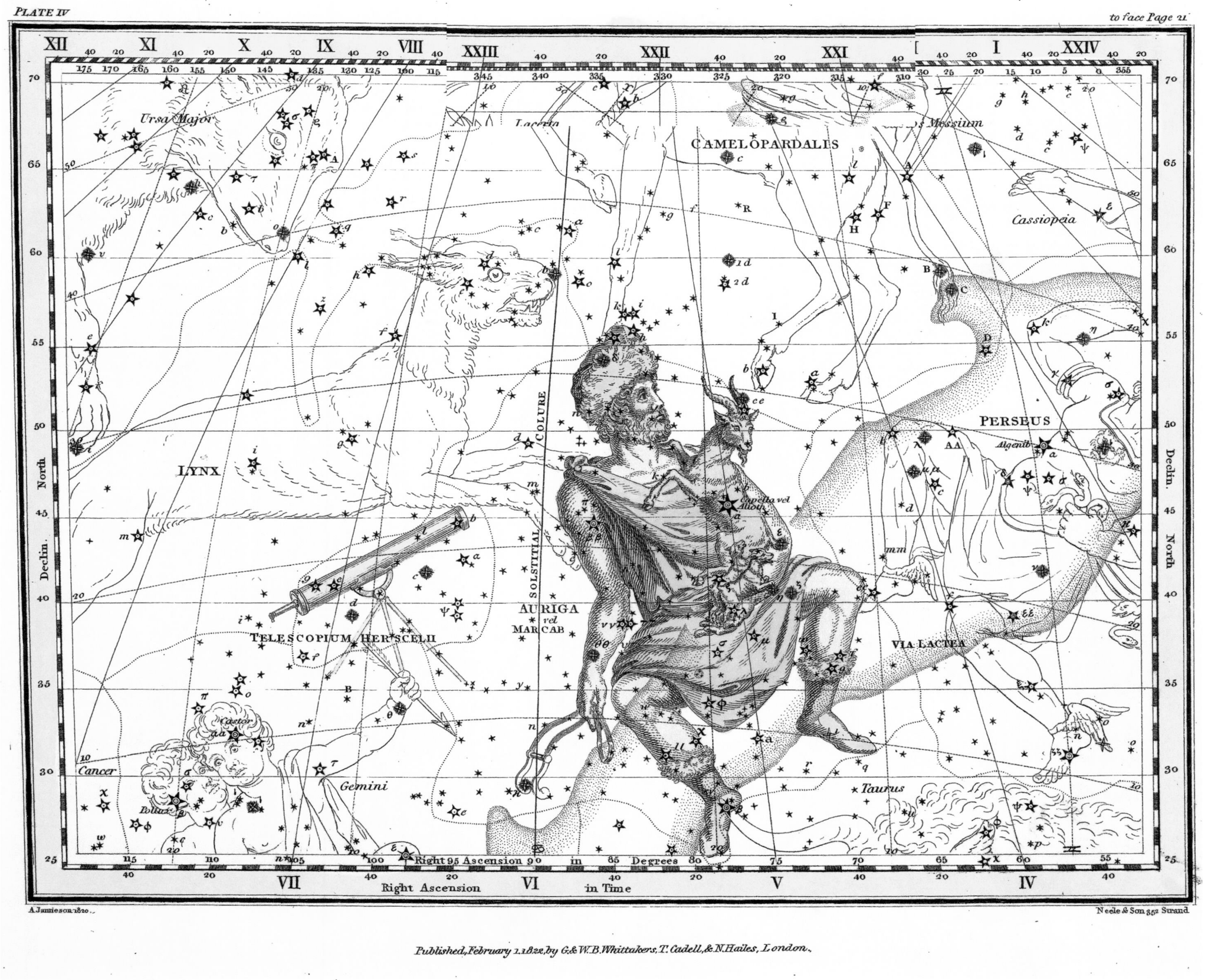

PLATE IV

to face Page 21

Published, February 1. 1822, by G. & W. B. Whittakers, T. Cadell, & N. Hailes, London.

Auriga, Camelopardalis, Telescopium Herschelii

One version of the Greek myth has it that the Auriga or Charioteer is Erichthonius, the legendary King of Athens and son of Hephaestus, the god of fire and volcanoes. He was raised by the goddess Athena and when he became king he yoked four horses to a chariot and even competed in the Panathenaic Games that he had founded in honor of the goddess. According to other legends, the charioteer is identified as Hippolytus, the son of Theseus, or with Myrtillus, the son of Hermes. The Auriga is holding the goat Amalthea, who suckled the newborn Zeus on the island of Crete: the god had the animal placed in the sky together with her two kid goats, and is the brightest star in the entire Auriga constellation, Capella (small female goat in Latin), while her two kids (Haedi) are the stars Eta and Zeta Aurigae. At left, the constellation that the astronomer Maximilian Hell dedicated to the telescope with which his colleague Frederick William Herschel discovered the planet Uranus.

Capricornus, Aquarius, Norma Nilotica

Here Jamieson represents the two zodiacal constellations of Capricorn and Aquarius. The latter is a bit ambiguous and difficult to interpret in that there are various legends concerning him. In some versions he is identified with Ganymede, the beautiful boy that Zeus carried away and transformed into the cupbearer for the gods; in others he is Deucalion, the son of Prometheus, who together with his wife Pyrrha survived the Flood (in this case he is depicted while pouring water). However, it is probable that the ancient Greeks merely produced a variation of an ancient Egyptian tradition according to which the constellation was the personification of the Nile. Jamieson introduced, in Aquarius' left hand, an alignment (now obsolete) he had created, the Norma Nilotica, an instrument used to measure the depth of the river.

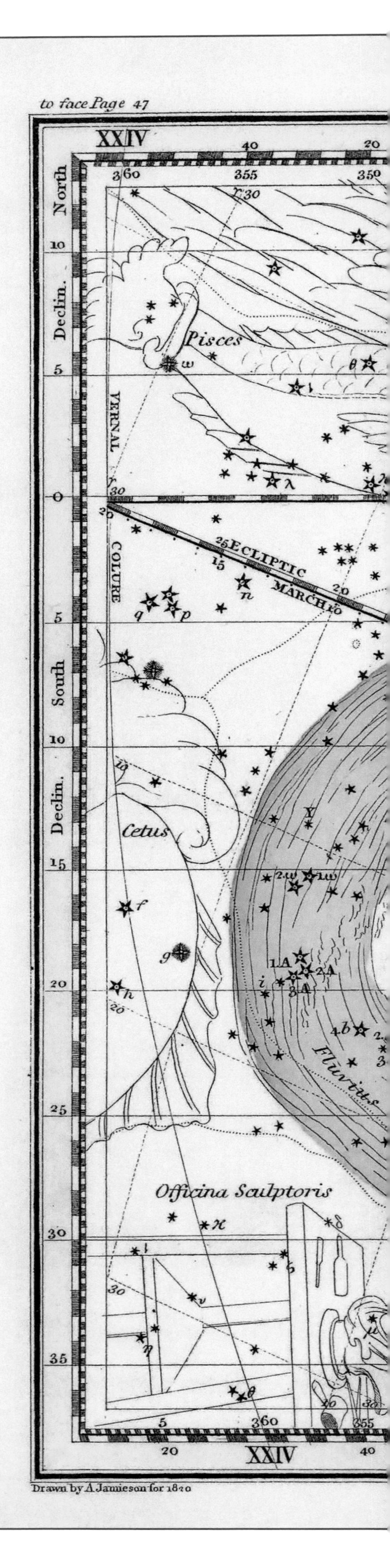

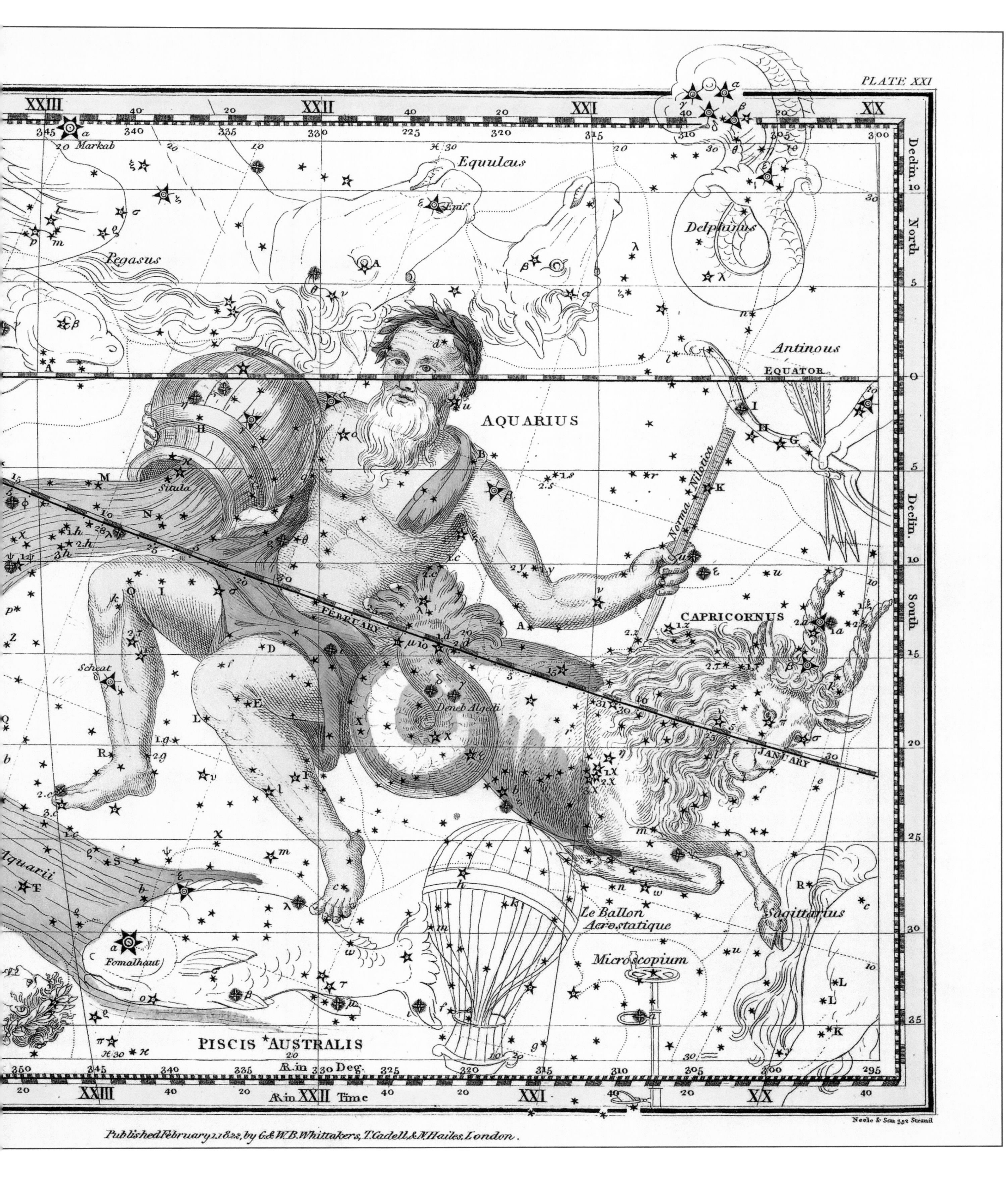
PLATE XXI
Markab
Equuleus
Delphinus
Pegasus
Antinous
EQUATOR
AQUARIUS
Situla
Norma Nilotica
CAPRICORNUS
FEBRUARY
JANUARY
Scheat
Deneb Algedi
Sagittarius
Le Ballon Aerostatique
Microscopium
Fomalhaut
Aquarii
PISCIS AUSTRALIS
Æ.in Deg.
Æ.in XXII Time
Declin. North
Declin. South
Neele & Son 352 Strand
Published February 1.1822, by G.&W.B. Whittakers, T.Cadell, & N.Hailes, London.

to face Page 30

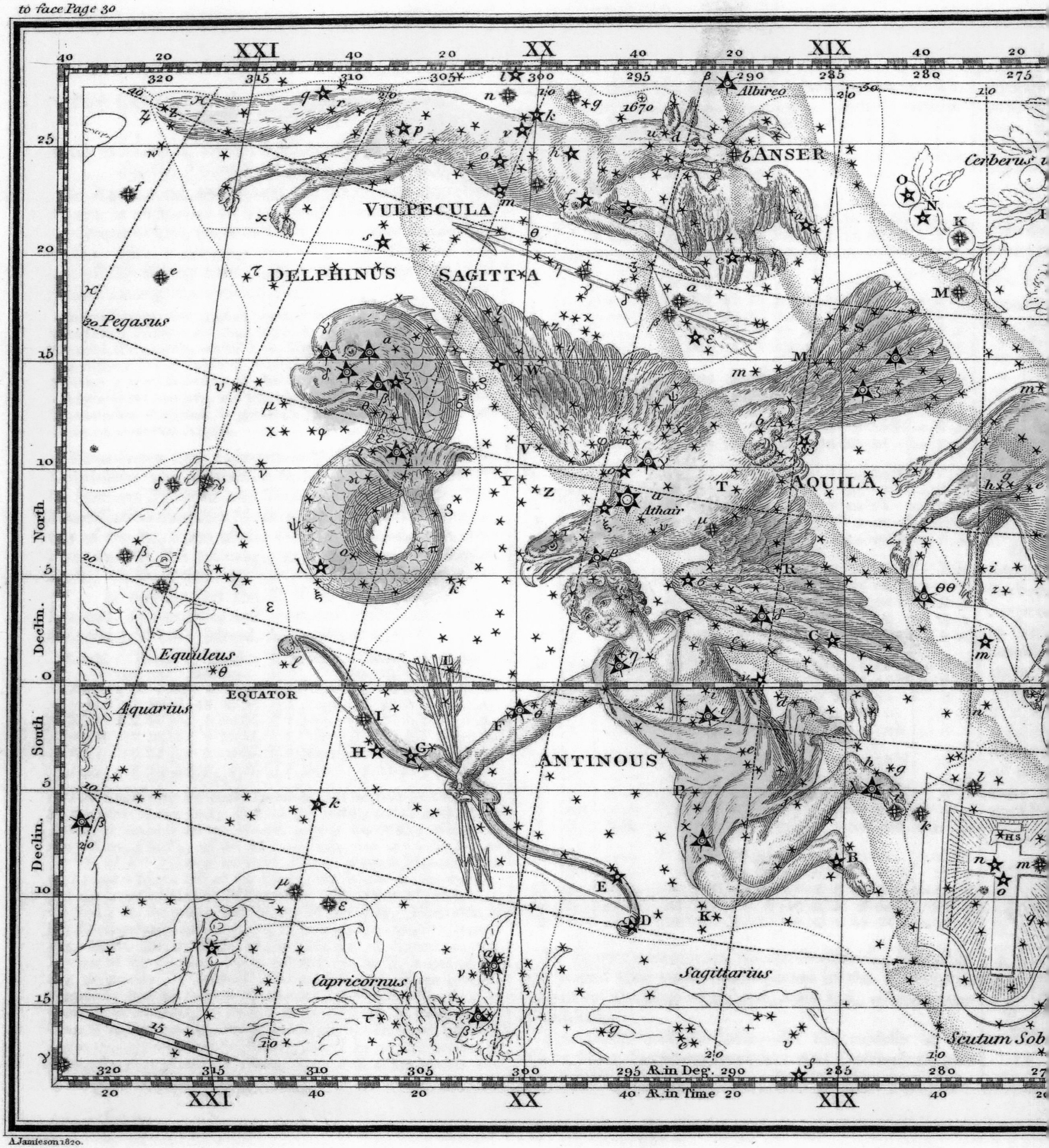

XXI XX XIX

VULPECULA

ANSER

Albireo

Cerberus

DELPHINUS

SAGITTA

Pegasus

AQUILA

Athair

Equuleus

EQUATOR

Aquarius

ANTINOUS

Capricornus

Sagittarius

Scutum Sob

North Declin.

South Declin.

Æ. in Deg.

Æ. in Time

A Jamieson 1820.

Published February 1.1822 by G&WB Whittakers T. Cadell & N. Hailes London.

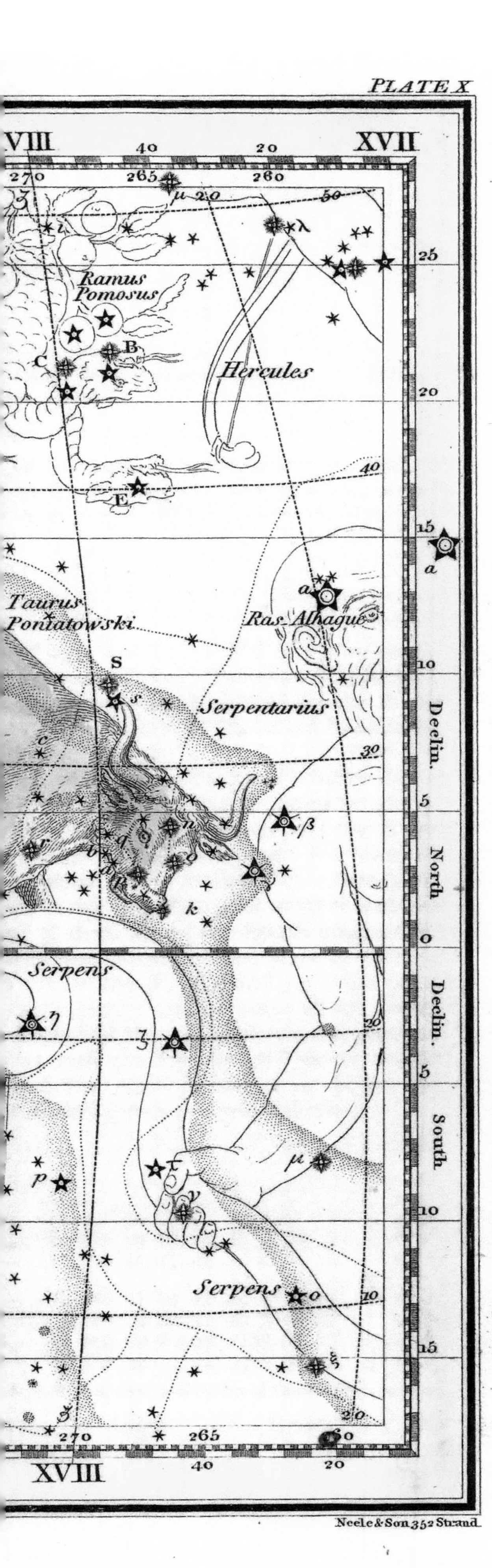

Aquila et Antinous, Scutum Sobieski, etc.

In the middle of this superb plate is the figure of Antinous, the handsome favorite of the Roman emperor Hadrian, who drowned in the Nile around AD 130. The first astronomer who named a constellation after him was Ptolemy, who wrote his Almagest *around twenty years after the youth's tragic demise. In celestial atlases Antinous is usually depicted as a boy being kidnapped by an eagle, the protagonist of the neighboring constellation. This is the reason why his figure is sometimes mistaken for that of the mythical Ganymede, who Zeus, in the likeness of an eagle, took to the sky. Jamieson depicts Antinous armed with bow and arrows. To his left (right, for the viewer) Jamieson re-introduces the modern constellation Scutum Sobiescianum, which here is called Scutum Sobieski or Shield of Sobieski.*

Leo

In Greek mythology, the zodiacal constellation of Leo represents the invincible lion that plagued the Argolic city of Nemea by attacking both its inhabitants and flocks. Hercules was ordered to kill him as the first of his Twelve Labors, but soon realized that his skin was wholly invulnerable. So the hero stunned the lion with his club and then used his incredible strength to strangle him with his bare hands. After he had killed the beast, Hercules skinned it by using the lion's claws and then wore the pelt as armor. The Nemean lion was transformed into a constellation by Zeus; its brightest star, on the animal's heart, is Regulus (Alpha Leonis), while the second brightest, Denebola (in Arabic, 'lion's tail'), is on its tail.

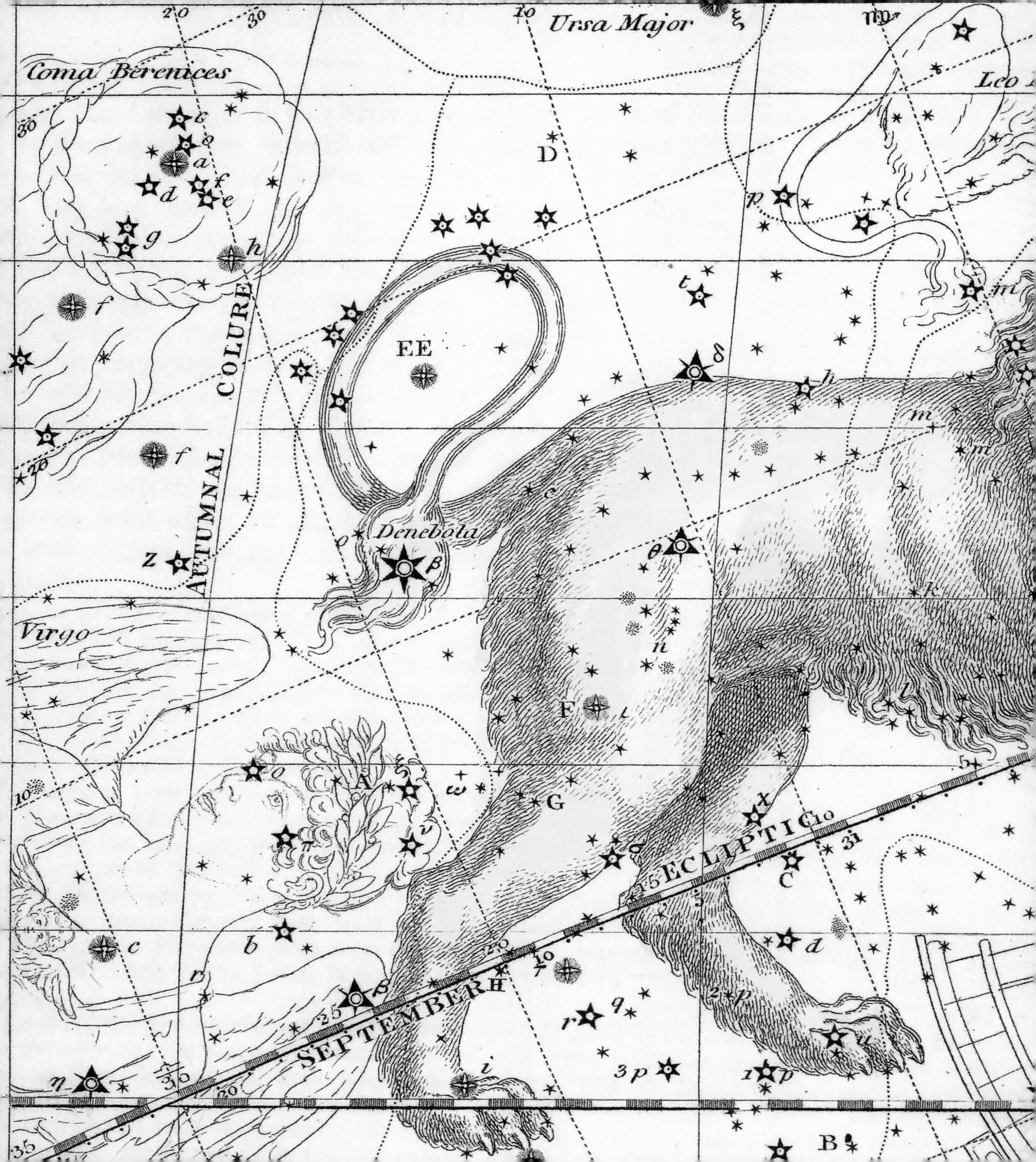

Ursa Major
Coma Berenices
Leo
D
EE
Denebola
AUTUMNAL COLURE
Virgo
F
G
A
C
ECLIPTIC
SEPTEMBER
B

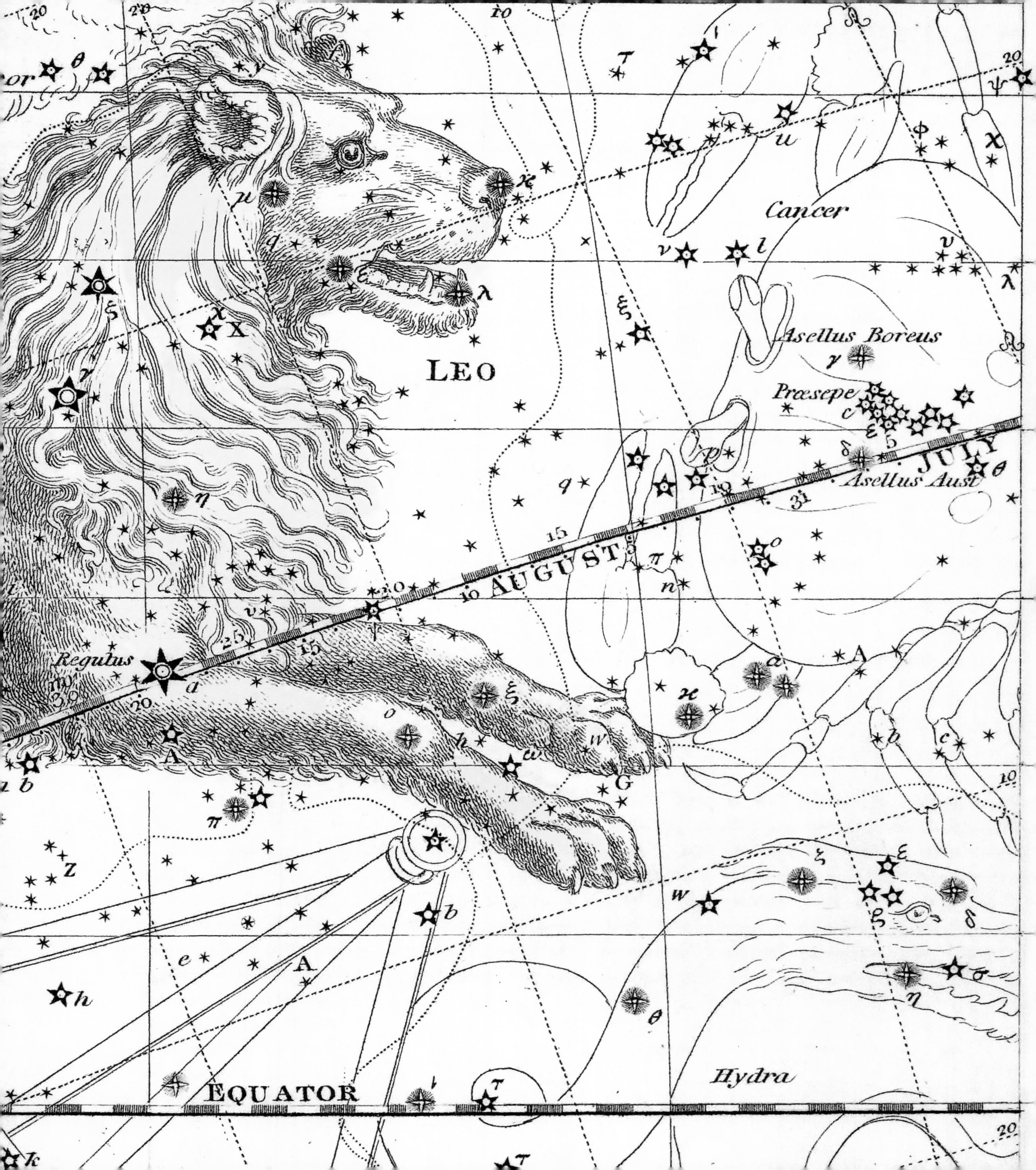
LEO
Cancer
Regulus
Asellus Boreus
Præsepe
Asellus Aust
AUGUST
JULY
EQUATOR
Hydra

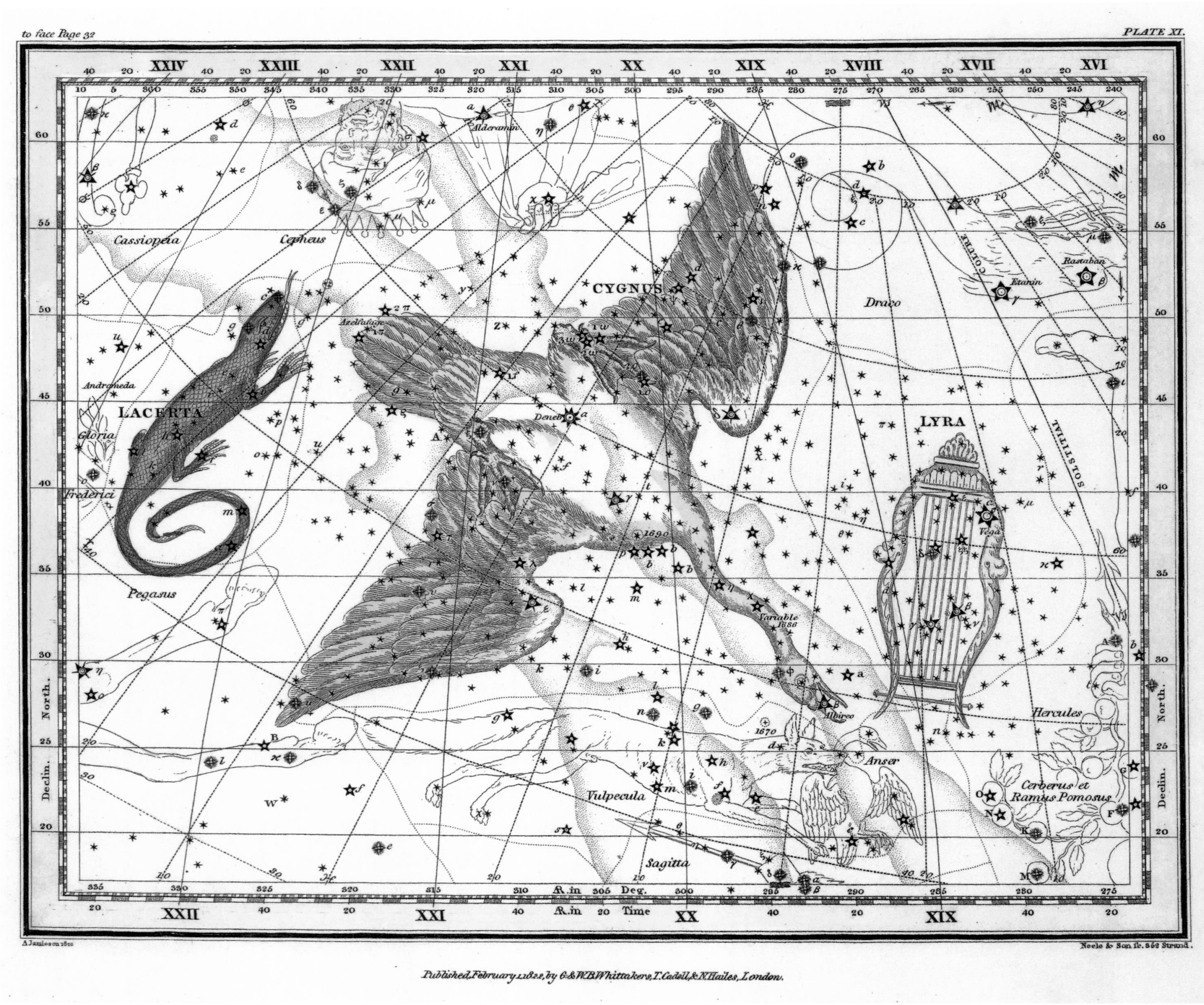

to face Page 32

PLATE XI.

A Jamieson 1820

Neele & Son Sc. 352 Strand.

Published February 1.1822, by G.&W.B.Whittakers, T.Cadell, & N.Hailes, London.

Cygnus, Lacerta, Lyra

Cygnus the Swan dominates an important part of the northern sky; its figure extends along the Milky Way in summer and is depicted while flying southward. Its brightest stars, which can easily be seen with the naked eye, form a cross-shaped asterism known as the Northern Cross. According to Greek mythology, the swan is one of the forms that Zeus took on to seduce one of his numerous lovers: Leda, the wife of King Tiyndareus of Sparta and the mother of Helen and the Dioscuri. The brightest star is Deneb, which usually lies along the Swan's tail (in this plate, however, it is in a more central position), while its beak is marked by Albireo. Next to the swan is the constellation Lyra, the most famous and brightest star of which is Vega. This latter, together with Deneb and Altair (which belongs to the Aquila constellation), form another major asterism, the Summer Triangle.

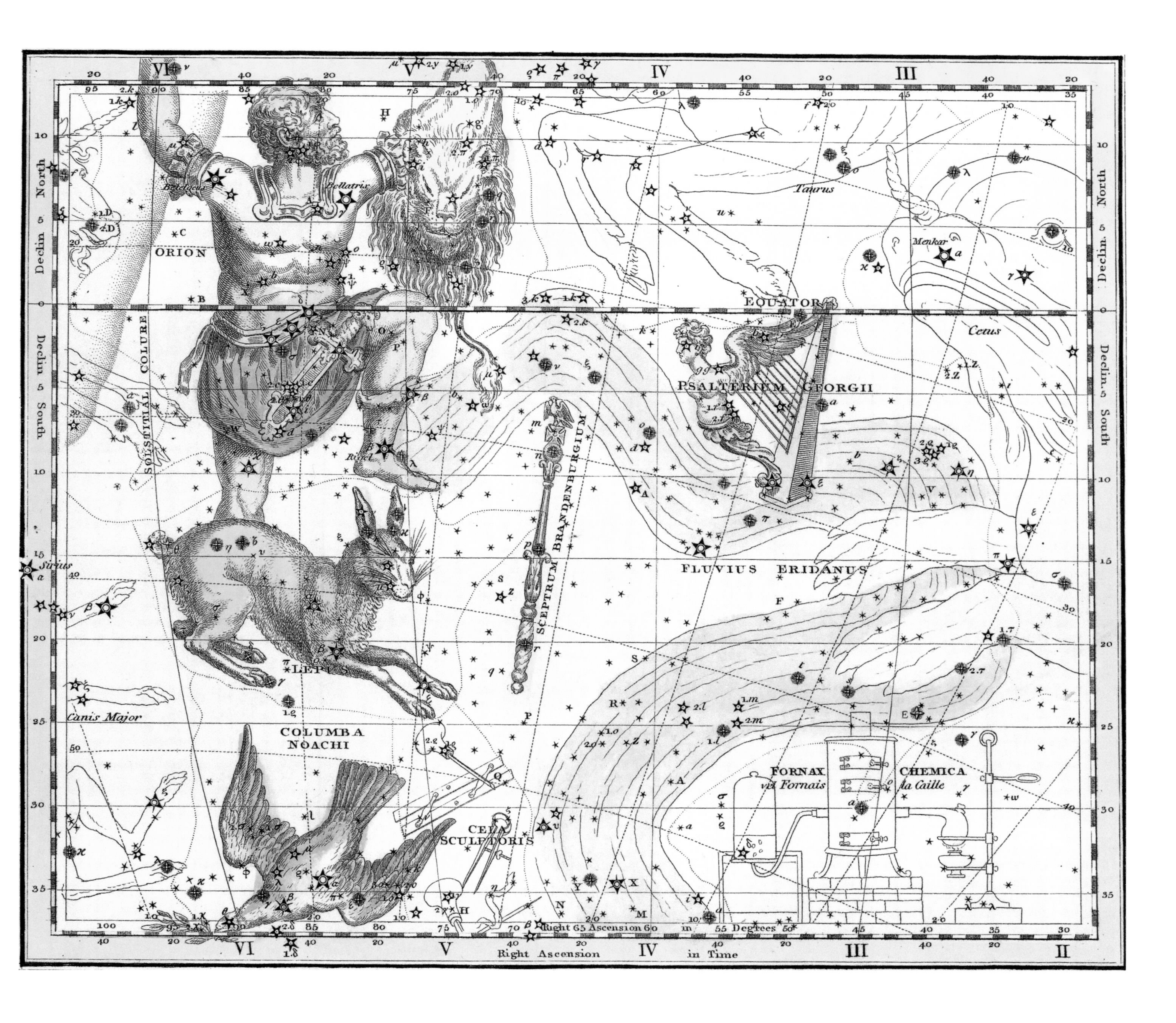

Eridanus, Orion, Lepus, Columba, Psalterium Georgii, Sceptrum Brandeburgium

In this plate Jamieson depicts Orion, Lepus (Hare), Columba (the Dove, which he calls Columba Noachi, or Noah's Dove), Eridanus, Psalterium Georgii (George's Harp) and Sceptrum Brandeburgium (which is really Sceptrum Brandenburgicum), the Brandenburg Specter constellation introduced in 1688 by the German astronomer Gottfried Kirch to honor the Prussian province of the same name, where he lived. Forgotten for almost a century, it was revived by Bode in 1782 and then in Uranographia *(1801), and is in turn also depicted by Jamieson here. This constellation is now obsolete and has been incorporated into the nearby Eridanus.*

Urania's Mirror

(1824)

In 1824, two years after the publication of *A Celestial Atlas, Comprising a Systematic Display of the Heavens* by Alexander Jamieson, the London publisher Samuel Leigh issued his *Urania's Mirror*, a set of 32 star chart cards, accompanied by a brief and simple treatise providing an explanation of astronomy: *A Familiar Treatise on Astronomy*, written by the American Jehoshaphat Aspin in a markedly 'popular', didactic fashion. The cards illustrate the 79 constellations visible in Great Britain and have holes in correspondence with the stars, which is an attractive way of allowing readers to see their position by placing the holes in front of a source of light. This brilliant idea – as well as its pocket format (c. 7.8 x 5.5 inches or 20 x 14 cm) and elegant box set decorated with a graceful female figure in the likeness of Urania, the Muse of astronomy – led to the huge popularity of the set among the public at large, which resulted in numerous editions.

However, *Urania's Mirror* almost immediately triggered a 'mystery'. The work was in all respects anonymous (except for the name of the engraver, Sidney Hall, which appears at the foot of each plate), but the drawings are practically identical to those in Jamieson's atlas. But if he was the author of the cards, why on earth would he have not used his name? And if, on the other hand, the illustrations were not his, who had executed them, and why?

Both the foreword to the set of cards and a passage from the accompanying treatise connect the design of the cards to a mysterious 'young woman', but neither her name nor biography was to be found. Some scholars believed that this strange attribution was nothing more or less than a stratagem on the part of the publisher to make astronomy more 'popular' and intriguing, since at the time this science was considered very difficult. As for the author, assuming it was a woman, Caroline Herschel and Mary Somerville, both talented and prominent astronomers in their day, were considered likely candidates; but there was no credible evidence that could confirm this. Thus, the doubts concerning the true identity of the author of *Urania's Mirror* remained obscure for more than 150 years, until in 1994, P.D. Hingley, the archivist of the Royal Astronomical Society, discovered the identity of the mysterious designer while re-arranging some old maps: a certain Reverend Richard Rouse Bloxam, very little of whom is known except for the fact that he was an assistant schoolmaster in the English town of Rugby. Why he wanted to remain anonymous is not certain, but it was probably because he wanted to avoid a possible accusation of plagiarism from Jamieson.

The Box Set

The 32 star chart cards of the constellations published under the title Urania's Mirror *were presented in a charming box, the lid of which was decorated with the figure of a young woman sitting on a cloud, holding a compass in her left hand and resting her right arm on the celestial sphere. The name of the publisher, Samuel Leigh of London, appears below the illustration, but no mention is made of the artist, whose identity was revealed only recently.*

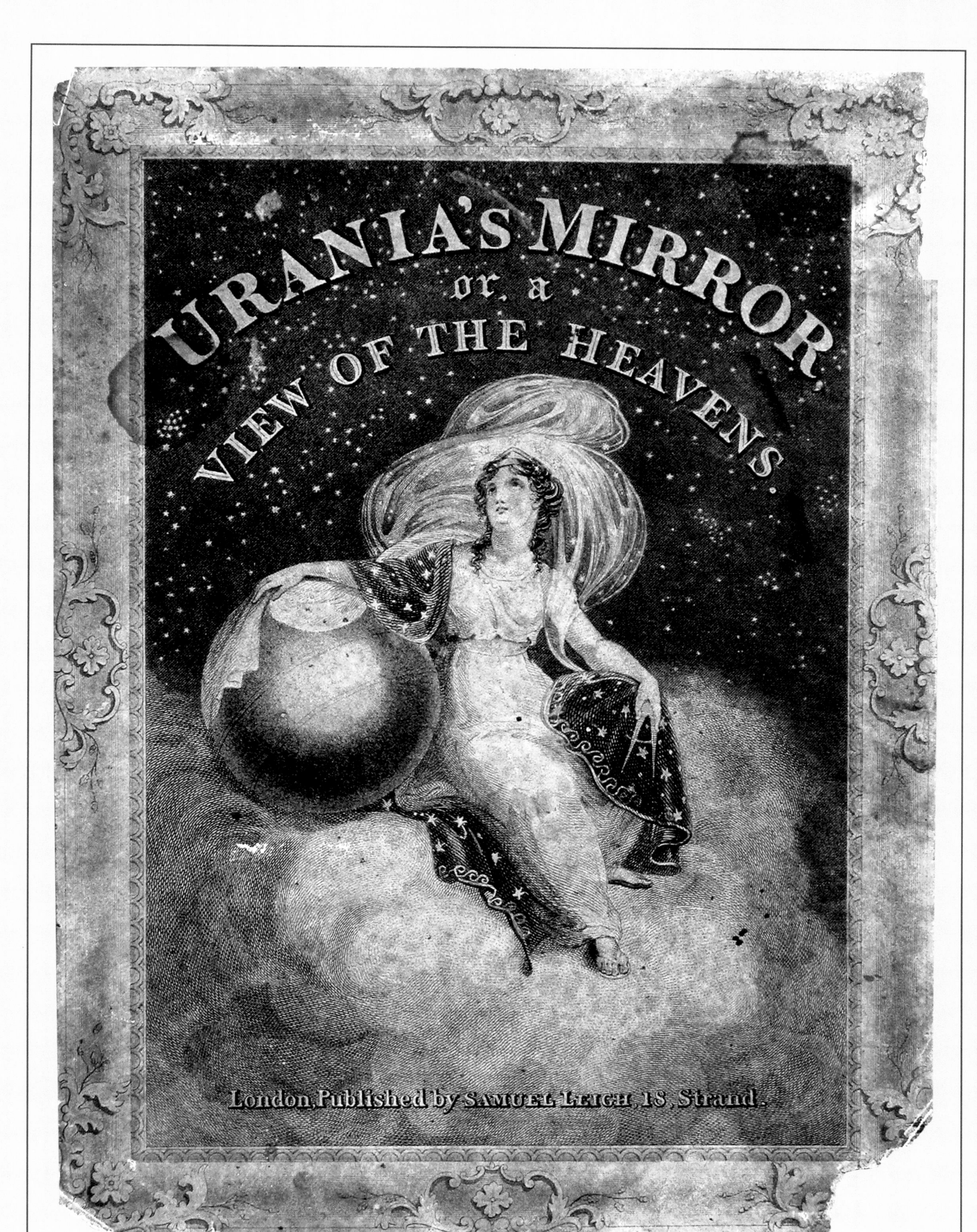
URANIA'S MIRROR
or, a
VIEW OF THE HEAVENS.
London, Published by SAMUEL LEIGH, 18, Strand.

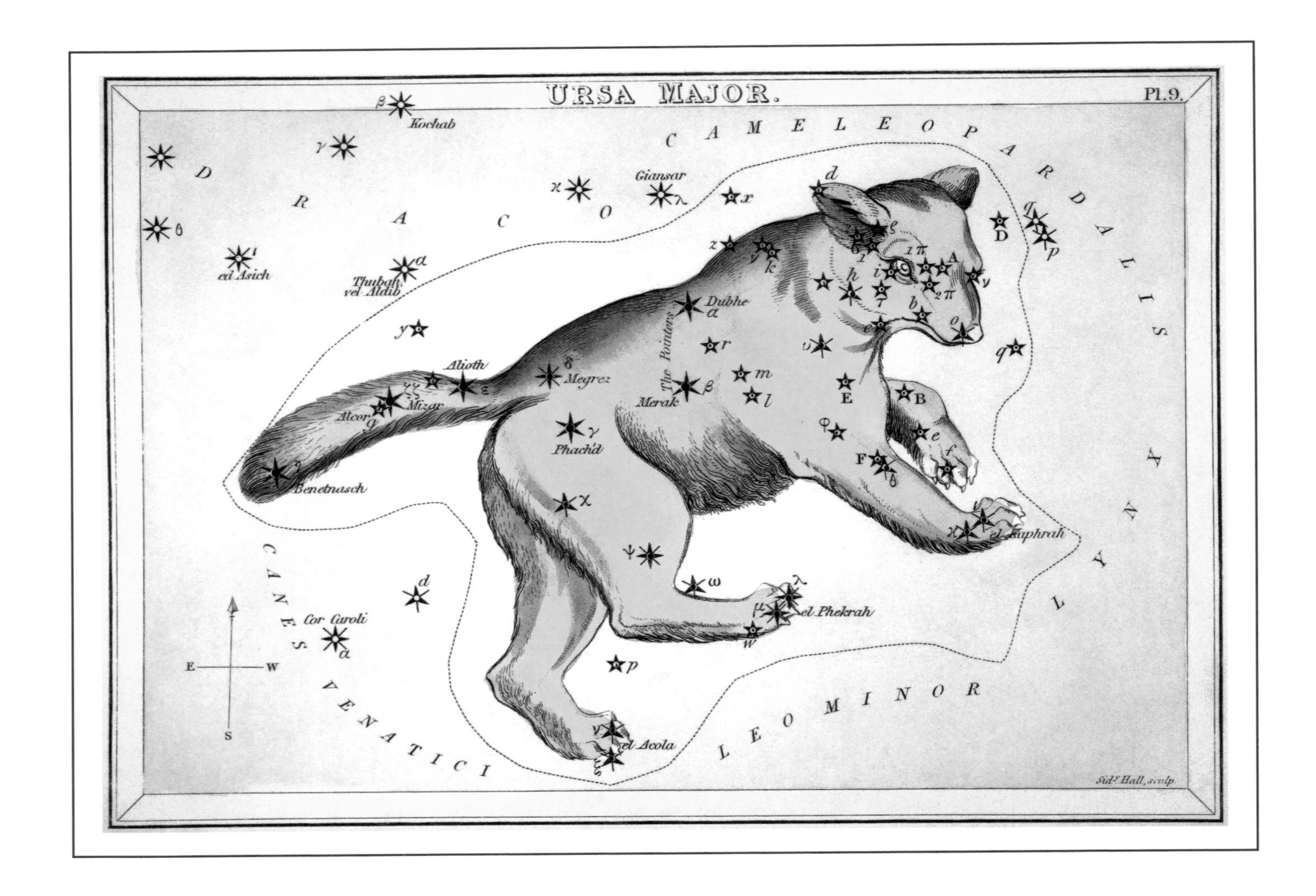

Ursa Major

The Great Bear dominates card number 9 of the collection. What is immediately visible to the viewer is the characteristic asterism made up of seven stars, in this order: Dubhe, Merak, Phecda, Megrez, Alioth, Mizar and Alkaid (here called Benetnash). The Romans called this star formation Septentrio (seven draught oxen), hence the word septentrional or 'northerly'. Near Mizar one can also see Alcor, which is barely visible to the naked eye. Furthermore, Merak and Dubh form another asterism known as Pointers since it 'points' toward Polaris, the North Star, which lies in the Ursa Minor constellation.

Draco and Ursa Minor

The set of star cards in Urania's Mirror *begins with the illustration of Draco and Ursa Minor (Dragon and Little Bear), on the tip of the tail of which is what is probably the most important star in the firmament: the North (or Pole) Star or Polaris (Alpha Ursae Minoris), which in the present-day precessional epoch indicates the celestial North Pole of Earth. The designer of the cards indicates Polaris with the alternative (and most ancient) names, Cynosura ('dog's tail' in Greek) and Alruccabah (in Arabic), which are just two of the many it has acquired. Ursa Minor has an asterism that very much reminds one of the Big Dipper in Ursa Major but is smaller and is therefore sometimes called Little Dipper.*

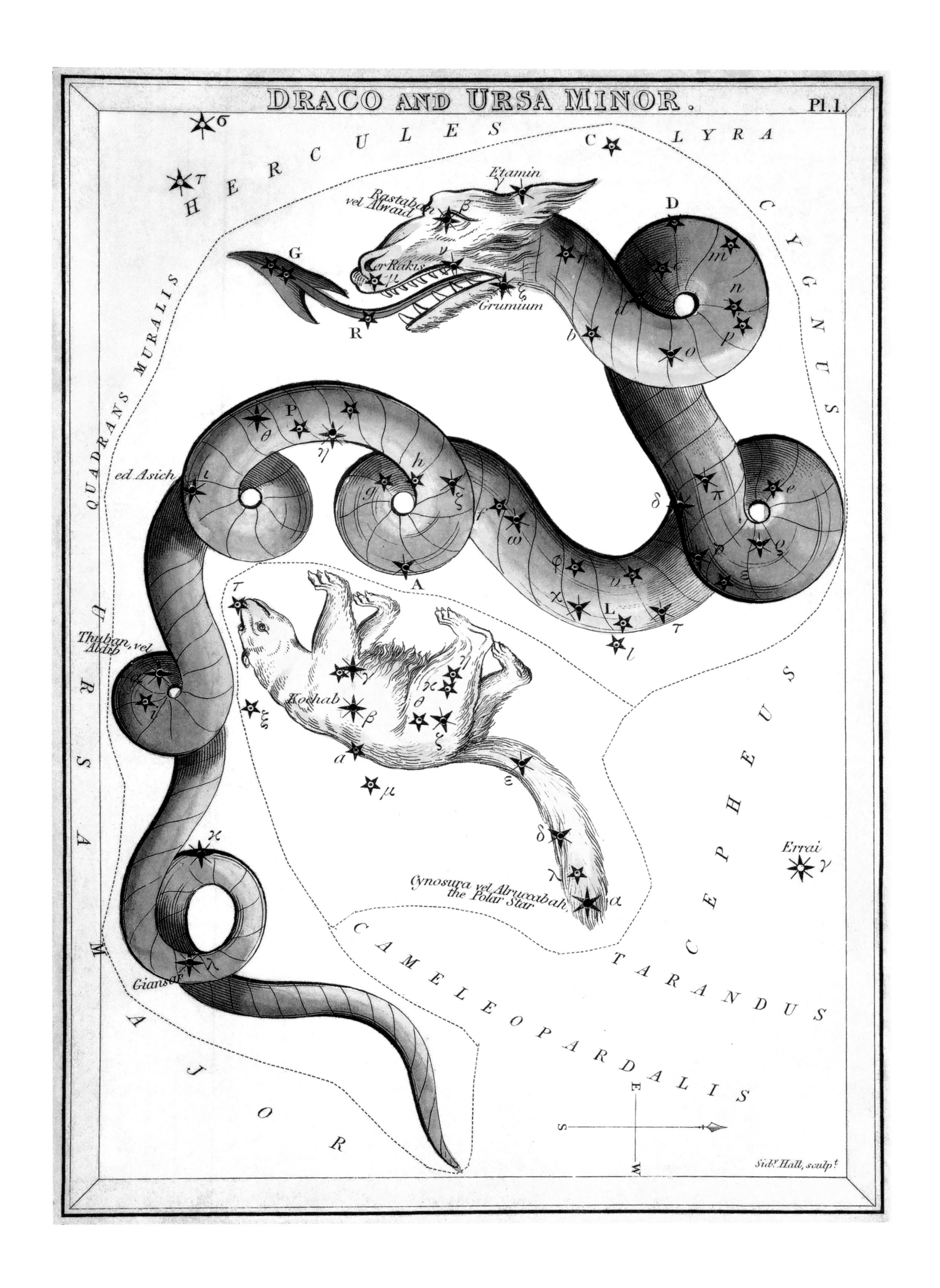
DRACO AND URSA MINOR.
Pl. 1.
HERCULES
LYRA
CYGNUS
QUADRANS MURALIS
URSA MAJOR
CEPHEUS
TARANDUS
CAMELEOPARDALIS
Etamin
Rastaban vel Alwaid
Er Rakis
Grumium
ed Asich
Thuban, vel Adib
Giansar
Kochab
Cynosura vel Alrucabah, the Polar Star
Errai
Sidy. Hall, sculpt.

Boötes, Canes Venatici, Comae Berenices, and Quadrans Muralis

Card number 10 represents the constellation of Boötes (the Plowman or Herdsman) together with those of the Hunting Dogs, Berenice's Hair and Mural Quadrant. The last-mentioned constellation, although now obsolete, is still important because the annual Quadrantid meteor shower, which radiates in this part of the sky every January, was named after it. It was introduced in the late 1700s by Joseph Jérôme de Lalande in honor of the quadrant mounted on the wall of the Paris Observatory (of which he was the director), and that allowed him to measure the position of the stars. In this illustration the Mural Quadrant is above Boötes' head: Bode had already placed it in his Uranographia, *but reduced its size compared to that established by Lalande.*

Virgo

The zodiacal constellation of Virgo (Virgin) occupies almost all the space in card number 21, as is only appropriate for the second largest constellation in the sky (the largest is Hydra). The brightest star is Spica (Ear of Grain), which is quite close to the celestial equator and is visible toward the west in summer after sunset. As its name indicates, this star is connected to the harvest period, and from an iconographic standpoint is here depicted in the middle of a bunch of ears of wheat. Another very important star is visible on the right arm of Virgo: Vindemiatrix (Grape Harvestress, or Epsilon Virginis); in ancient times it rose at dawn toward the end of August and was therefore associated with the imminent grape harvest.

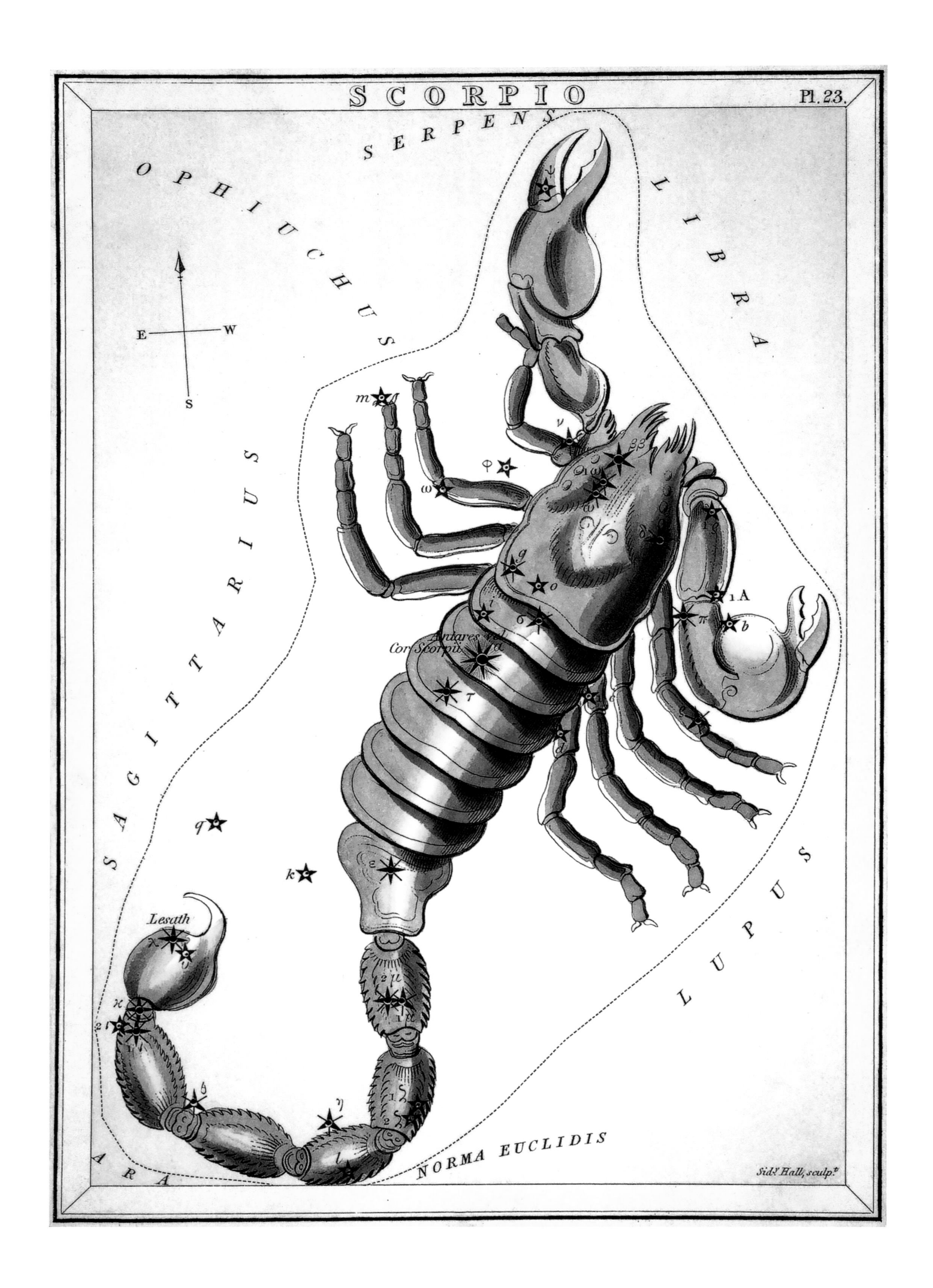
SCORPIO
Pl. 23.
SERPENS
OPHIUCHUS
LIBRA
E
W
S
SAGITTARIUS
Antares vel
Cor Scorpii
Lesath
LUPUS
NORMA EUCLIDIS
ARA
Sid.y Hall, sculp.t

Scorpio

Card number 23 is dedicated to the zodiacal constellation Scorpion. The stars are in the same position and have the same notation as those in Jamieson's analogous representation. The star Lesath (Upsilon Scorpii), the eighth brightest star in the constellation, is located in the animal's stinger, as is the neighboring Shaula (Arabic for 'stinger'), which is the second brightest star after Antares (which lies in the heart) but is merely called Lambda Scorpii. Despite the fact that the latter is the twenty-fourth brightest star in the sky, Bayer classified it as 'lambda', the eleventh letter of the Greek alphabet, perhaps because it is located in a very southward position in the sky.

Aries and Musca Borealis

The Ram and Northern Fly are reproduced in card number 16. The former is a well-known zodiacal constellation, while the latter, now obsolete, was one of the star formations conceived by the Dutch-Flemish astronomer Petrus Plancius and illustrated in the celestial globe he made in 1612. The story of its name is quite complicated: it was originally Apis (Bee) and was renamed Vespa (Wasp) in 1624 by the German astronomer Jakob Bartsch, and in 1690 Hevelius gave it yet another name, Musca (Fly), in his Firmamentum Sobiescianum. *But here the constellation ran the risk of being mistaken for an analogous one already in the southern sky, so later astronomers decided to change the name to Musca Borealis. Again, in 1674 Ignace-Gaston Pardies named it Lilium (Lily or Fleur-de-lis) but a short time later this name was definitively discarded.*

TAURUS
AURIGA
PERS
El Nath
GEMINI
Aldebaran
HY
Meissa
Betelgeux
Bellatrix
ORION
Orion's Belt
ERIDANUS
E
W
S

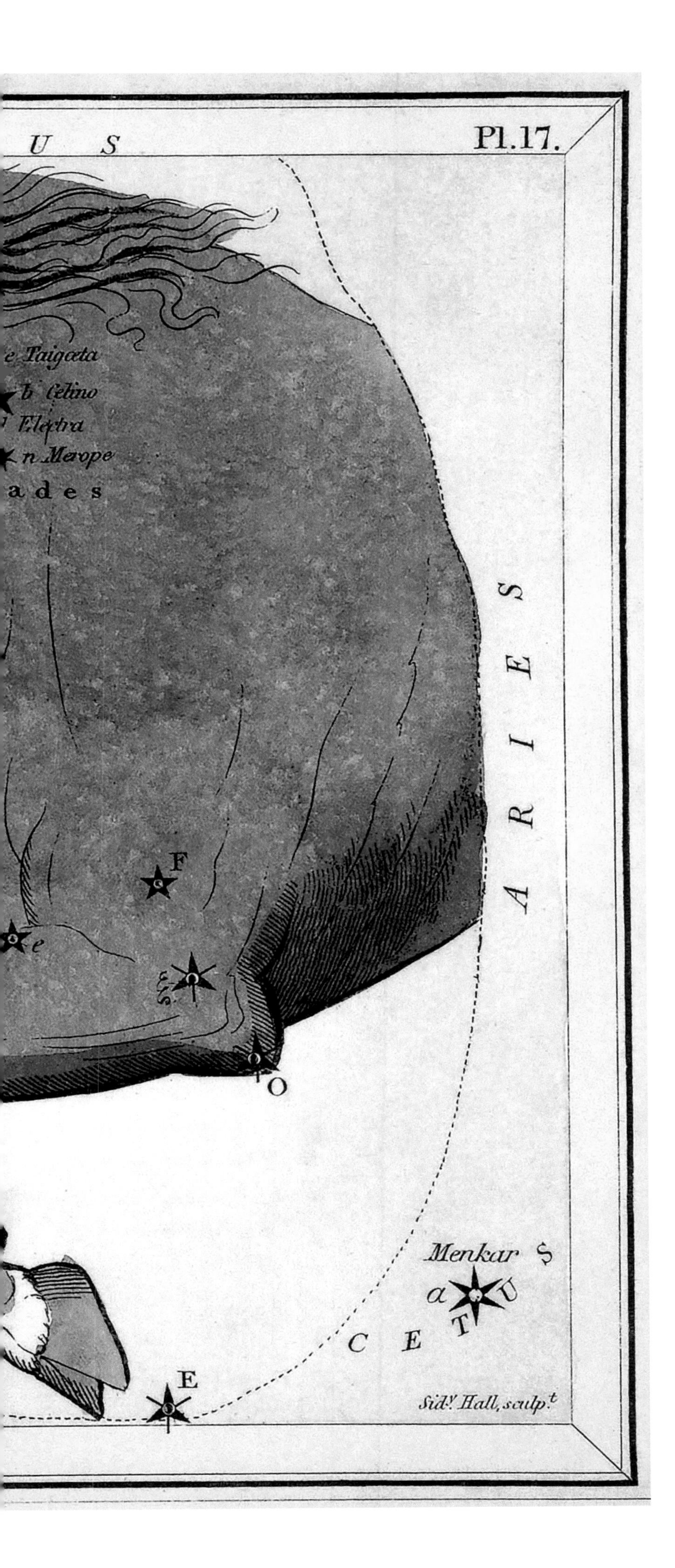

Taurus

In the card dedicated to the Bull constellation (number 17) only the animal is illustrated, accompanied by the stars. Standing out in particular are the Pleiades, an open cluster of stars that have been considered extremely important since ancient times because of their luminosity, which makes them visible to the naked eye. They were named after the mythical seven sisters, daughters of Atlas and Pleione born in Arcadia: Alcyone, Celaeno, Electra, Maia, Merope, Sterope (or Asterope) and Taygete. According to one of the many versions of the Greek myth, the girls were continuously pursued by the hunter Orion, who forced them to run away continuously. One day the gods took pity on the sisters and transformed them into doves (in fact, this is the meaning of their name); they later became stars.

Monoceros, Canis Minor, and Atelier Typographique

The Unicorn, Little Dog and Printer's Workshop (here with its French name, Atelier Typographique) are illustrated in card number 31, the penultimate of the series. The rendering of Printer's Workshop in this illustration is quite simplified (for example, Jamieson had placed a realistic rendering of a printing press at right) and the star formations are also very schematic. In fact, in Canis Minor the author chose to indicate only the principal stars (Procyon and Gomeisa, which here is called Gomelza), ignoring those of lesser magnitude. Furthermore, this card has only the names of the neighboring constellations and of their most important stars, for example Betelgeuse (here called Betelgeux), the second brightest star in nearby Orion (the brightest is Rigel, which is not illustrated because it lies outside the card).

NOCEROS, CANIS MINOR.
Pl. 31.
GEMINI
Gomelz
Procyon
Betelguex
ORION
Alnitak
Saiph
CANIS MAJOR
LEPUS
Muliphen
var.
Sirius
Mirzam
Sid.y Hall, sculp.t
ATELIER TYPOGRAPHIQUE.

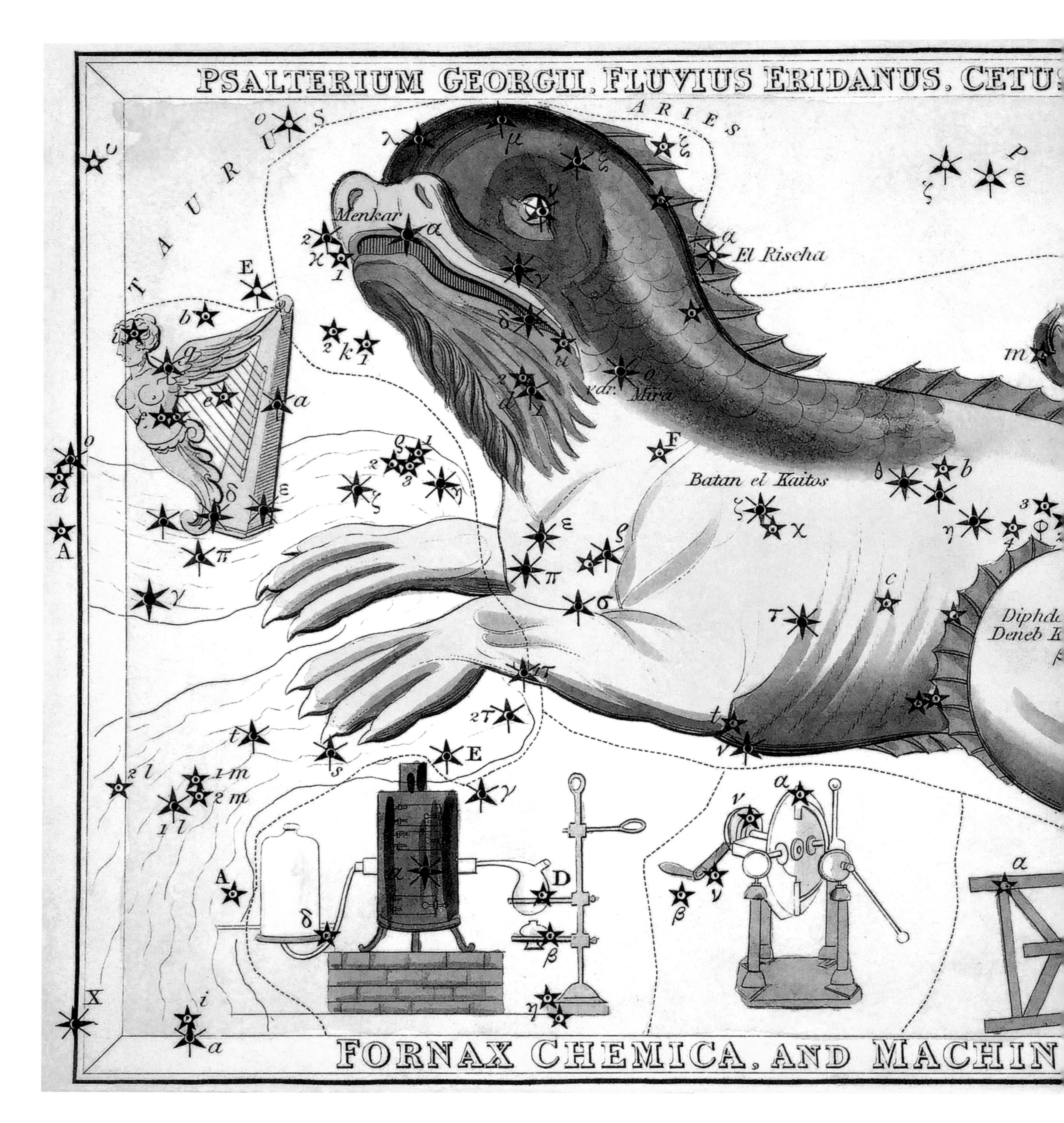
PSALTERIUM GEORGII, FLUVIUS ERIDANUS, CETUS
FORNAX CHEMICA, AND MACHIN
ARIES
TAURUS
Menkar
El Rischa
var. Mira
Batan el Kaitos
Diphda
Deneb K

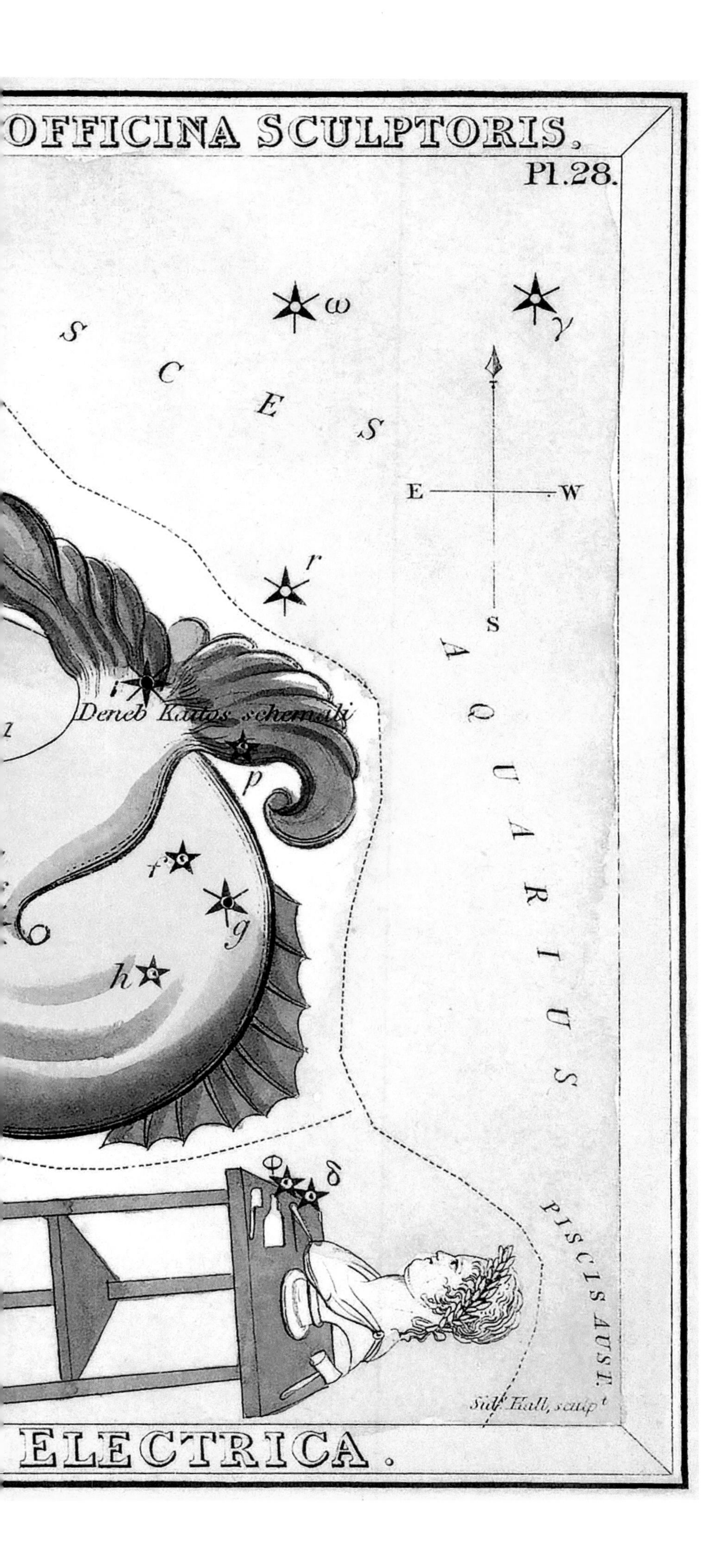

Psalterium Georgii, Fluvius Eridanus, Cetus, Officina Sculptoris, Fornax Chemica, and Machina Electrica

Card number 28 represents the group of southern sky constellations in the sector dominated by Cetus (Whale or Sea Monster): the Eridanus River, Sculptor (or Sculptor's Studio), George's Harp, Chemical Furnace and Electric Generator (now obsolete). The principal stars of Cetus are indicated by their names: Deneb Kaitos (Beta Ceti, in the middle of the tail), Menkar (Alpha Ceti, on the muzzle), Baten Kaitos (here called Batal el-Kaitos, in the middle of the body), and so on. With regard not only to Cetus, but all the constellations in this card, their position and the rendering of their respective illustrations are quite similar to Jamieson's analogous illustrations.

CANIS MAJOR, LEPUS, COLUMBA NOACHI &
MONOCEROS
ORI
ATELIER TYPOGRAPHIQUE
ARGO NAVIS
Sirius
Mirzam
Muliphen
Wesen
Adhara
Aludra

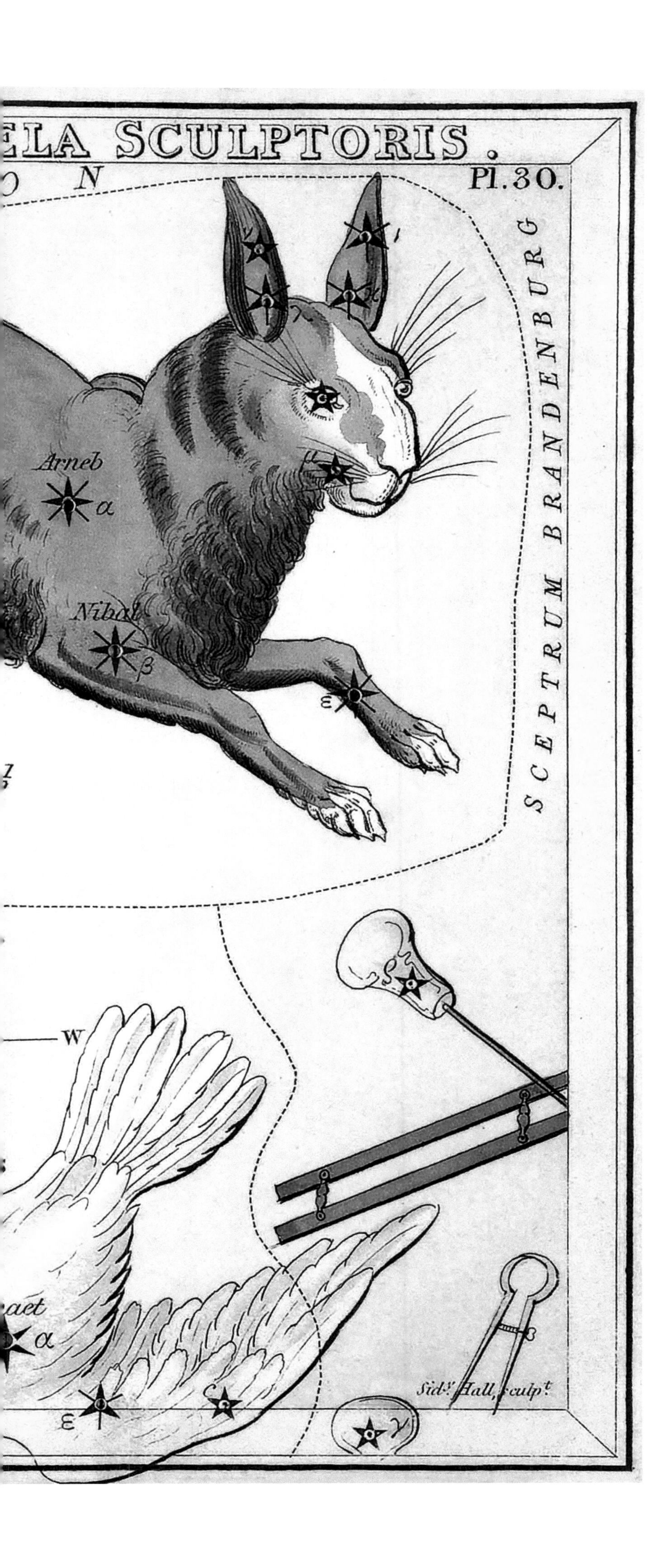

Canis Major, Lepus, Columba Noachi & Cela Sculptoris

This plate, featuring the northern sky constellations Greater Dog, Hare, Dove and Engraver's Chisel (now Caelum or Chisel), is the thirtieth of the collection. As regards Greater Dog, the illustrator places the brightest star, Sirius, on the muzzle. Hare, which theoretically should be pursued by the Dog, is here portrayed in a rather static pose; however, both the brightest stars – Arneb and Nihal – are easily seen and accompanied by their names in full. Unlike the other two constellations, Cela Sculptoris was introduced in 1756 by Lacaille and is the eighth smallest in the firmament. In the illustration it is only partially seen in the lower right, accompanied by a compass (without stars).

Noctua, Corvus, Crater, Sextans Uraniae, Hydra, Felis, Lupus, Centaurus, Antlia Pneumatica, Argo Navis, and Pyxis Nautica

The set ends with a flourish. The last card, number 32, depicts a large group of southern sky constellations, including two introduced by Lacaille: Antlia Pneumatica (Pneumatic Machine or Air Pump) – inserted here in honor of the device invented by the French scientist Denis Papin to recreate a vacuum in the laboratory – and Pyxis Nautica (Mariner's Compass). Jamieson is to be credited for introducing Noctua, the night owl, as a replacement of Lemonnier's Solitarius. Lastly, here Sextant is indicated with its complete name (as it was in Doppelmayr's atlas): Sextans Uraniae (Urania's, or Astronomical, Sextant), representing the instrument used by Hevelius (who introduced this constellation) in his observations of the sky.

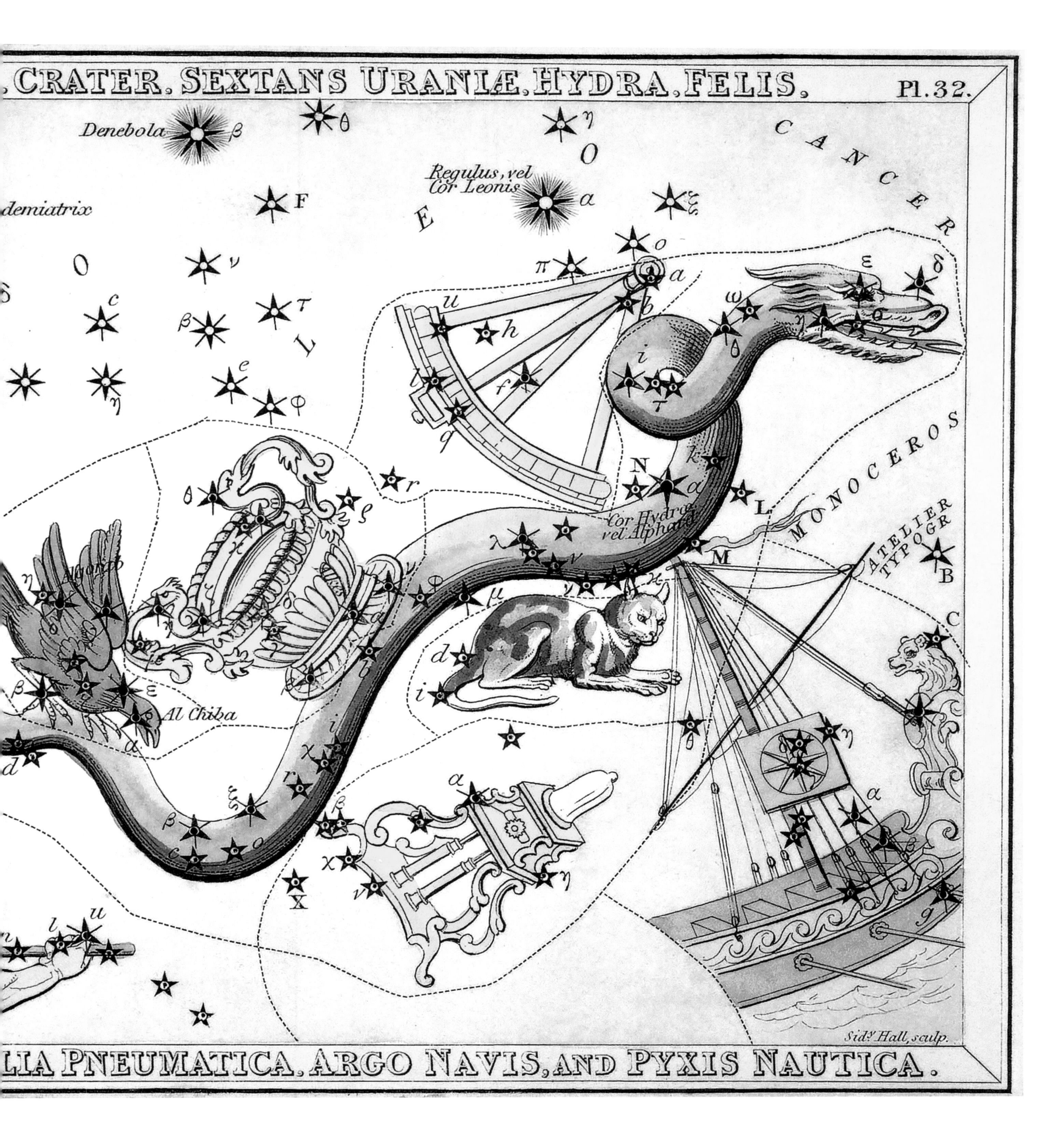
CRATER, SEXTANS URANIÆ, HYDRA, FELIS,
Pl. 32.
Denebola
Regulus, vel
Cor Leonis
demiatrix
CANCER
MONOCEROS
ATELIER
TYPOGR
Cor Hydræ
vel Alphard
Al Chiba
Sidy. Hall, sculp.
LIA PNEUMATICA, ARGO NAVIS, AND PYXIS NAUTICA.

Elijah Hinsdale Burritt

(1794-1838)

The American Elijah Hinsdale Burritt was the classic example of a self-made man in his country. He came from a family of immigrants from Wales that had become poor after the flour mill they managed went bankrupt, yet despite this he managed to become the pioneer of American astronomy. The eldest of ten children, Burritt was born on 20 April 1794 in New Britain, Connecticut. After acquiring a scrappy education at public schools, at the age of eighteen he was sent to Simsbury, where he became a blacksmith's apprentice. However, a serious accident crippled him for some time and he could not continue this line of work. Of a naturally curious character and gifted with brilliant intelligence, Burritt had become passionate about astronomy at an early age and profited from his enforced inactivity to study this science. His friends were impressed by his passion and capabilities, and as soon as his condition improved they helped him to enroll at Williams College, which he attended regularly, while at the same time giving lessons in order to pay for his studies. But he never earned a degree.

In 1819 Burritt moved to Milledgeville, at that time the capital of Georgia, in the Deep South, where he married Ann W. Watson (they would have five children) and found employment in the technical offices of the state, where he was charged with drawing up a study of the course of the Chattahoochee River, which is part of the Georgia-Alabama border. This was the period of the furious debate between the defenders of slavery, most of whom were concentrated in the southern states, and the abolitionists, who were active in the north of the United States. Although Burritt did not openly support the abolitionists' stance, neither did he bother to hide their booklets and pamphlets, which his friends from the North sent him regularly. Quite the contrary, he left them – perhaps on purpose – scattered around his office. It didn't take much time for people to notice this, which triggered suspicions about and reactions to his beliefs that became so virulent that in 1829 he was forced to leave Georgia as quickly as possible, abandoning all his property. He returned to New Britain and had to start from scratch. But he didn't lost heart: he moved in his parent's house, part of which he converted into an observatory; he purchased a telescope and with the aid of his brother Elihu opened a school; and he devoted much energy to writing works on arithmetic and astronomy.

Four years later Burritt published his most important work, Geography of the Heavens, *which, although conceived as a fundamentally didactic book (it was intended as a class book for students), immediately became a basic point of reference in the burgeoning American astronomy circles. It contained a description of the sky, including the seasonal variations in the two hemispheres, in eight plates, while the constellations were presented by means of the traditional illustrations, the only difference being that only the brightest stars were represented.*

In 1837 Burritt organized a group of thirty colonists – including his sister Emily and another brother, William – who set out to start life anew in the Republic of Texas, which had just become independent from Mexico. However, the journey proved to be extremely difficult and ill-fated for the group. After 28 days of navigation, when their ship was about to land at Galveston it was struck by a violent storm. The group survived miraculously and was able to resume the trip, but once it had arrived at Houston the members were unable to find any accommodation and were forced to stay in an improvised tent camp. Overwhelmed by hardship, many members of the party caught yellow fever, including Burritt himself, who died on 3 January 1838 when only 44 years old, just when his companions had finally succeeded in arriving at their destination.

Northern Circumpolar Map

This map represents the circumpolar constellations of the northern sky visible every month of the year. The magnitude of the individual stars can be seen in the legend in the lower left of the map. Note that most of the constellations (such as Ursa Major, Ursa Minor, Cygnus and Auriga) are indicated both by their traditional Latin name and their corresponding English translation.

PL.VI.

NORTHERN CIRCUMPOLAR MAP for each Month in the Year.

NOVEMBER
DECEMBER
JANUARY
FEBRUARY
MARCH
APRIL
MAY
JUNE
JULY
AUGUST
SEPTEMBER
OCTOBER

GLORIA FREDERICA
ANDROMEDA
CASSIOPEIA R.12.D.60.
PERSEUS R.46.D.49.
Cygnus THE SWAN
CEPHEUS R.338.D.68.
CAMELOPARDALIS R.68.D.70.
Auriga THE WAGGONER R.75.D.45.
FIRST MERIDIAN
SOLSTITIAL COLURE
Polar Star
Draco THE DRAGON R.270.D.65.
Ursa Minor THE LESSER BEAR R.225.D.75.
THE LYNX
Ursa Major THE GREATER BEAR R.163.D.50.
THE DIPPER
EQUINOCTIAL COLURE
HERCULES
MURAL QUADRANT
BOOTES

Scale of Magnitudes
1 2 3 4 5 6 Cl. Nb.

Engraved by W. G. Evans New York, under the Direction of E. H. Burritt.

Hartford, Published by F. J. HUNTINGTON 1835, Entered according to act of Congress Sept. 1st 1832, by F. J. Huntington, of the State of Connecticut.

Geography of the Heavens

(1833)

Geography of the Heavens was the last great illustrated atlas of the 19th century. Harking back to Bode's tradition, Burritt had intended to title this work *Uranography*, but his publisher, Huntington and Savage, felt the title was too difficult for the public of school and university students who were its target, and so the two opted for something more simple and immediate. The work is divided into two parts. The first, which is theoretical, contains a description of the constellations and their history, illustrating the most important stars, including their magnitude and basic features. This is followed by a section dedicated to the solar system, with all the known planets, their satellites and comets. Lastly, there is a set of problems and exercises that the students were to solve with the help of thirteen detailed astronomical tables. The second part of the volume contains the celestial atlas proper, with illustrations of the constellations in keeping with the traditional figurative approach. The maps, eight in all, were drawn by Burritt himself, engraved under his direct supervision, and then painted with vivid colors.

The two-part volume came out in 1833 sold out in a short time, so that a second edition was published two years later, which enjoyed such an enormous success that the third edition came out the following year, with the addition of an introduction by the Scotsman Thomas Dick, a distinguished author of scientific works, and a ninth map illustrating the Sun, Moon and planets, including their respective sizes, distances and orbits. A fourth edition, with larger maps, was printed in 1841, when Burritt was already dead. Other editions of *Geography of the Heavens* were issued the following years, making it one of the classics of American popular astronomy.

Southern Circumpolar Map

This map illustrates the circumpolar constellations (those that are always above the horizon at a fixed latitude) of the southern sky each month of the year. Of these, Carina (or Keel, once part of the Argo Navis constellation), Centaur and Crux or Southern Cross are always visible from any given latitude of the Southern Hemisphere, while the others appear at certain times of the year and then 'set', depending on the location from which they are observed. The stars are indicated according to the scale of magnitude in the lower left-hand section of the map.

Pl. VII.
SOUTHERN CIRCUMPOLAR MAP
for each Month in the Year.
NOVEMBER
OCTOBER
SEPTEMBER
AUGUST
JULY
JUNE
MAY
APRIL
MARCH
FEBRUARY
JANUARY
DECEMBER
THE PHENIX
Horologium, THE CLOCK
Toucana, THE AMERICAN GOOSE
Achernar
Solarium, THE SUN DIAL
HYDRUS
The Water Snake
R. 28. D. 66.
Pavo, THE PEACOCK
R. 302. D. 68.
Telescopium THE TELESCOPE
Dorado, SWORD FISH
R. 75. D. 62.
Equuleus Pictoria THE PAINTERS EASEL
Hadleys Quadrant OCTANS HADLEIANUS
R. 310. D. 80.
SOLSTITIAL
COLURE
MERIDIAN
Canopus
Piscis Volans, THE FLYING FISH
R. 127. D. 68.
THE CAMELIAN
APUS, BIRD of PARADISE
R. 252. D. 7.
SOUTHERN TRIANGLE
R. 270. D. 65.
Ara, THE ALTAR
Argo Navis THE SHIP
Robur Caroli KING CHARLES'S OAK
Musca Indica, INDIAN FLY
R. 185. D. 69.
Circinus, THE COMPASSES
Lupus, THE HARE
Acrux
Crux, THE CROSS
Agena
EQUINOCTIAL COLURE
THE CENTAUR
R. 200. D. 50.
Scale of Magnitudes
1 2 3 4 5 6 Cl. Neb.
Engraved by W. G. Evans N. York, under the Direction of E. H. Burritt.
Hartford, Published by F. J. HUNTINGTON, 1835, Entered according to act of Congress Sept. 1st 1835, by F. J. Huntington, of the State of Connecticut.

Pl.VIII.

A CELESTIAL PLANISPHERE, OR MAP OF THE

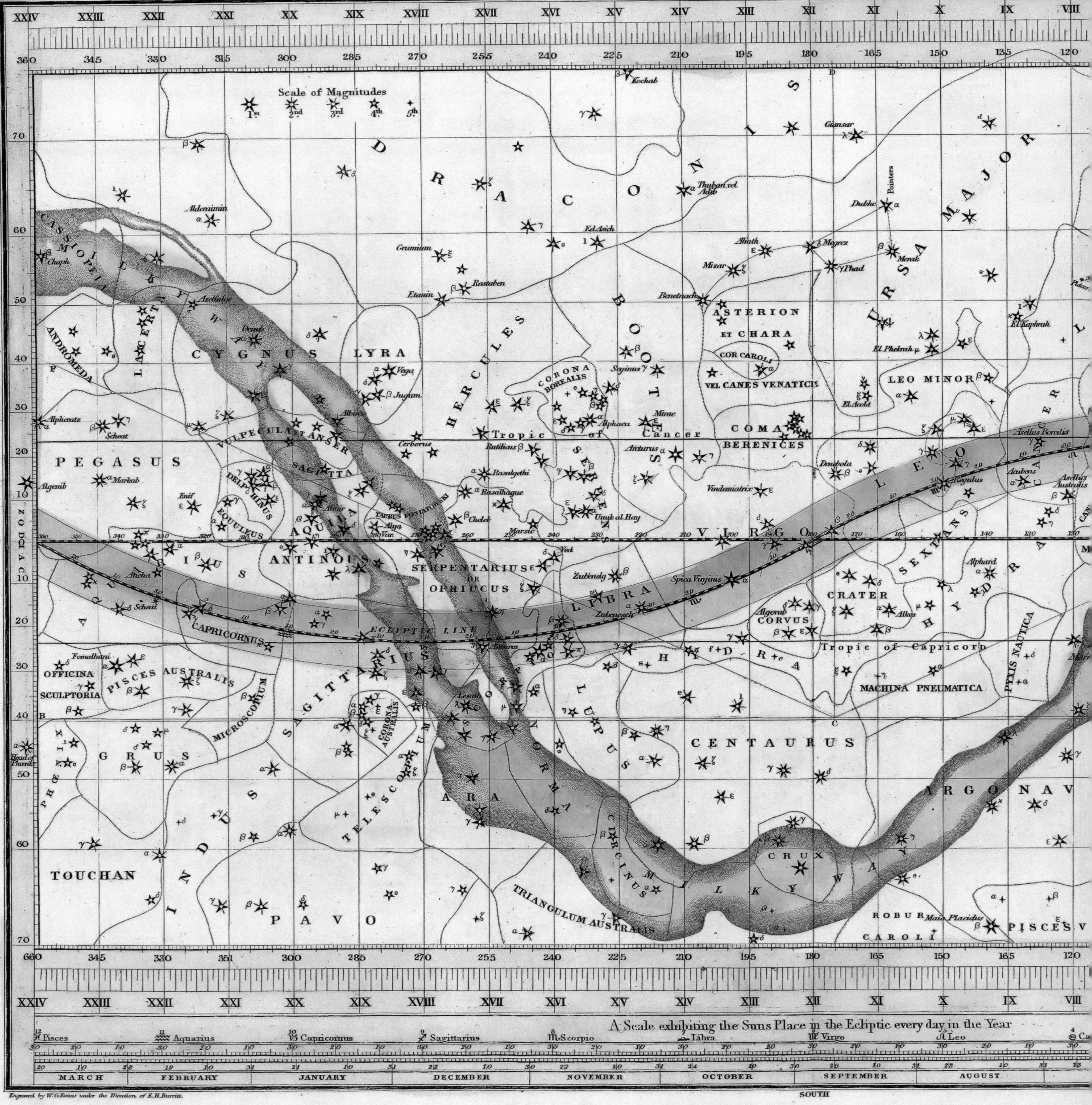

XXIV XXIII XXII XXI XX XIX XVIII XVII XVI XV XIV XIII XII XI X IX VIII

Scale of Magnitudes 1st 2nd 3rd 4th 5th

A Scale exhibiting the Suns Place in the Ecliptic every day in the Year

Pisces · Aquarius · Capricornus · Sagittarius · Scorpio · Libra · Virgo · Leo · Ca

MARCH · FEBRUARY · JANUARY · DECEMBER · NOVEMBER · OCTOBER · SEPTEMBER · AUGUST

SOUTH

Engraved by W. G. Evans under the Direction of E. H. Burritt.

Hartford, Published by F. J. HUNTINGTON 1835, Entered according to act of Congress Sept. 1st. 1835 by F. J. Huntington of the State of Connecticut.

A Celestial Planisphere, or Map of the Heavens

Burritt's work is for the most part based on those of Doppelmayr and Bode. This map illustrates the constellations on the planisphere, with their respective stars (only the principal ones) depicted differently, depending on their magnitude, which is shown in the scale in the upper left. Furthermore, the map is traversed by the Milky Way, the galaxy to which our solar system belongs. Lastly, the lower margin of the chart has a calendar indicating the position of the Sun every day of the year.

September, August, July / June, May, April

Pages 202-203: these two maps illustrate the visible sky divided into quarters. At left, the summer constellations (July-September) and at right, the spring ones (April-June).

March, February, January / December, November, October

Pages 204-205: the constellations visible in the winter sky (left, January-March) and autumn sky (right, October-December).

Hartford. Published by F. J. HUNTINGTON 1835 Entered according to act of Congress Jan.y 1 1835 by F. J. Huntington of the State of Connecticut.

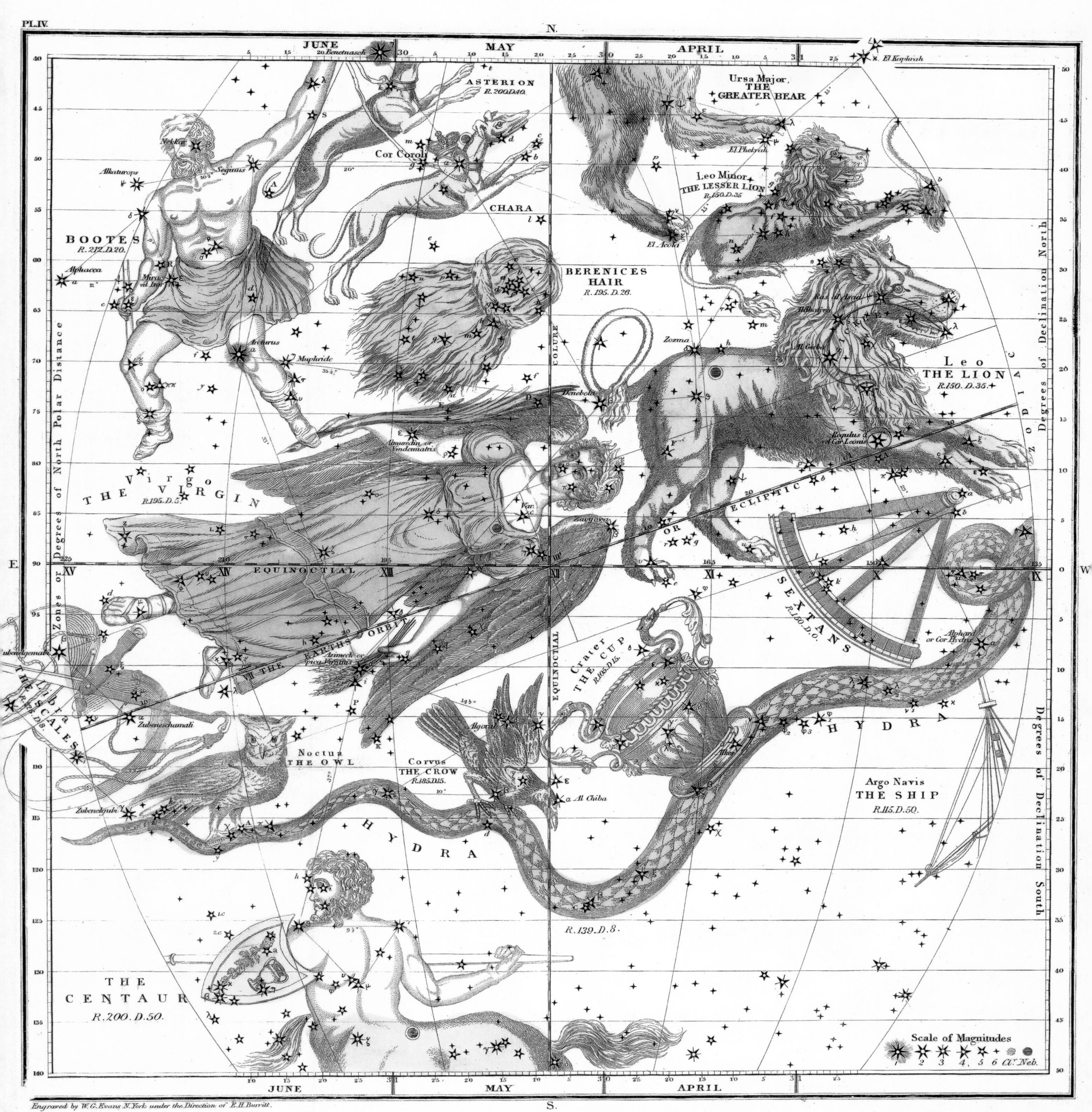
PL.IV.
N.
JUNE
MAY
APRIL
20 Benetnasch
ASTERION
R.200.D.40.
Ursa Major
THE
GREATER BEAR
El Kophrah
El Phekrah
Nekkar
Seginus
Alkaturops
Cor Caroli
CHARA
BOOTES
R.212.D.20.
Alphacca
Mirac
Arcturus
Mufride
BERENICES
HAIR
R.195.D.26.
Leo Minor
THE LESSER LION
R.150.D.35.
El Acola
Zosma
Denebola
COLURE
Leo
THE LION
R.150.D.35.
Regulus or Cor Leonis
ZODIAC
ECLIPTIC
Vindemiatrix
THE VIRGIN
Virgo
R.195.D.5.
Zavijava
EQUINOCTIAL
E.
W.
SEXTANS
R.150.D.0.
THE EARTHS ORBIT
Spica Virginis
Libra
THE SCALES
Zubeneschamali
Zubenelgubi
Crater
THE CUP
R.165.D.15.
Alphard or Cor Hydrae
HYDRA
Noctua
THE OWL
Corvus
THE CROW
R.185.D.15.
Algorab
Al Chiba
Alkes
Argo Navis
THE SHIP
R.115.D.50.
R.139.D.8.
THE
CENTAUR
R.200.D.50.
Scale of Magnitudes
1 2 3 4 5 6 Cl. Neb.
Zones or Degrees of North Polar Distance
Degrees of Declination North
Degrees of Declination South
S.
Engraved by W.G. Evans N. York under the Direction of E.H. Burritt.
Hartford Published by F.J. HUNTINGTON 1835, Entered according to act of Congress Sep. 1st 1835 by F.J. Huntington of the State of Connecticut

Engraved by W. G. Evans N. York under the Direction of E. H. Burritt.

Hartford, Published by F. J. HUNTINGTON 1835. Entered according to act of Congress Sept. 1st 1835, by F. J. Huntington, of the State of Connecticut.

PL. II.

N.

DECEMBER NOVEMBER OCTOBER

Caput Medusæ, HEAD OF MEDUSA

Algol

Almaach

Triangula, THE TRIANGLES

Musca, THE FLY
R. 40. D. 28.

Aries, THE RAM

Arietis or El Nath

Sheratan

Mesarthim

Pisces Borealis, THE NORTHERN FISH

Mirach

Andromeda
R. 14. D. 30.

Delta

Alpheratz

Gloria Frederica
FREDERICK'S GLORY

Lacerta, THE LIZARD

THE SWAN

Cygni

Pegasus, THE FLYING HORSE
R. 340. D. 14.

The Square of Pegasus

Scheat

Markab

Algenib

Enif

Equuleus
R. 316. D. 5.

Kitel Phard

THE PISCES FISHES
R. 5. D. 10.

ZODIAC

THE EARTHS ORBIT

EQUINOCTIAL

Piscis Occidentalis, THE WESTERN FISH

OR ECLIPTIC

EQUINOCTIAL COLURE

FIRST MERIDIAN OF THE HEAVENS

Menkar

El Rischa

Cetus, THE WHALE
R. 25. D. 12.

Baten Kaitos

Deneb Kaitos Schemali

Diphda or Deneb Kaitos

Aquarius, THE WATER BEARER

Sad El Melik

Sad Es Saud

Ancha

Situla

Scheat

Fluvius Aquarii, RIVER AQUARIUS or the Cascade

Capricornus, THE SEA GOAT

Deneb Algedi

Piscis Australis, THE SOUTH FISH
R. 335. D. 30.

Fomalhaut

THE PHŒNIX
R. 10. D. 50.

Grus, THE CRANE
R. 330. D. 45.

Scale of Magnitudes
1 2 3 4 5 6 Cl.r Neb.

Zones or Degrees of North Polar Distance

Degrees of Declination North

Degrees of Declination South

E. W.

DECEMBER NOVEMBER OCTOBER

S.

Engraved by W. G. Evans N. York, under the Direction of E. H. Burritt.

Hartford, Published by F. J. HUNTINGTON 1835, Entered according to act of Congress Sept.r 1st 1835, by F. J. Huntington, of the State of Connecticut.

The Author

*E*LENA **PERCIVALDI**, *(Milan, 1973) is a medievalist and essayist. She has been a professional journalist since 2002, collaborating with some of the high-quality history periodicals:* Medioevo, BBC History, Storie di Guerre e Guerrieri, *and* Conoscere la Storia. *She is also editor of* Storie & Archeostorie, *a history, art and archaeology newsletter published by Perceval Archeostoria, the research and consultancy studio headed by Percivaldi herself. She has also been very active as a speaker in meetings, congresses and the like throughout Italy and abroad, and has organized exhibitions concerning history and archaeology as well as historic-commemorative events. She is a member of prestigious institutions, as well as of the Advisory Committee of the periodical* Medioevo Italiano. *Elena Percivaldi earned a degree at Milan University with a thesis on medieval history with the transcription of almost 200 parchments, most of which were still unpublished, concerning the vicarage of Santo Stefano Church in Vimercate. Since that time she has continued working in this vein, always affirming the importance of documentary sources as an indispensable basis of any historic research. A passionate scholar of ancient manuscripts and books, she was a consultant for the CAPSA Ars Scriptoria publishing project for a facsimile edition of certain major Longobard codices. Furthermore, in 2008 she edited and translated the first complete edition – including an introduction, commentary, and the translation in Italian with the Latin text en face (and with a preface by Franco Cardini) – of* Navigatio Sancti Brandani *(a 9th-10th century Irish text), for which she was awarded the 2009 Edition of the Premio Italia Medievale. She has also been involved in lectures, articles and publications concerning the sky, the calendar and the cosmology of the past, as well as with other fascinating aspects of ancient and medieval history. Among her books are the following:* I Celti. Una civiltà europea; I Lombardi che fecero l'impresa. La Lega e il Barbarossa tra storia e leggenda; Fu vero Editto? Costantino e il Cristianesimo tra storia e leggenda; La vita segreta del Medioevo; Gli Antipapi. Storia e segreti.

Acknowledgements

Writing a book on such a vast, difficult, and extremely fascinating subject as celestial atlases in history is a task that is anything but simple. Indeed, this subject is very complex, and the numerous available sources (beginning with the editions of many atlases that I have covered, as well as the scientific literature that studies it in detail) are scattered in dozens of libraries all over the world. Fortunately, many of these libraries place their manuscripts and printed books at the disposal of the public online. Thus I have been able to virtually leaf through (but with the same emotion!) the precious copies thanks to the digital versions realized by institutions such as the BEIC (Biblioteca Europea di Informazione e Cultura), the Bibliothèque Nationale de France, the University of Vienna, the Linda Hall Library, the Museo Galileo (brunelleschi.imss.fi.it), and by portals such as e-rara.ch (a platform for the consultation of old editions that have been digitalized and kept in Swiss libraries) and Certissima Signa (certissimasigna.sns.it), created as part of a research project on illustrated astronomical manuscripts of the Scuola Normale Superiore of Pisa. Another indispensable documentary source was Felice Stoppa's *Atlas Coelestis* web site (www.atlas-coelestis.com), an extraordinary 'compass' with which I was able to find my way about in the vast literature in this field. It is complemented by the text of the same name, *Atlas Coelestis. Il cielo stellato nella scienza e nell'arte* (*Celestial Atlas. The Starry Sky in Science and Art*), written by the same author and published by Salviati Editore. Last in order but first in importance, I want to thank my parents, my husband Mario, and my sons Riccardo and Jacopo for the moral support they gave me while engaged in both the research and actual writing. Without them I would not have been able to produce this book.

Photo credits

Page 5 Courtesy of the David Rumsey Map Collection, www.davidrumsey.com
Page 7 British Library, London, UK/Bridgeman Images
Page 8 British Library, London, UK/Bridgeman Images
Page 9 De Agostini/Bridgeman Images
Pages 10-11 Leemage/Universal Images Group/Getty Images
Page 13 SSPL/Getty Images
Page 15 British Library, London, UK/Bridgeman Images
Page 16 De Agostini/G. Cigolini/Getty Images
Page 17 Fine Art Images/Heritage Images/Getty Images
Page 18 Leemage/UIG/Getty Images
Page 19 De Agostini Picture Library/Getty Images
Page 21 Courtesy of the David Rumsey Map Collection, www.davidrumsey.com
Page 23 Jay M. Pasachoff/Science Faction/Getty Images
Pages 24-25 Jay M. Pasachoff/Science Faction/Getty Images
Pages 26-27 Jay M. Pasachoff/Science Faction/Getty Images
Page 28 Jay M. Pasachoff/Science Faction/Getty Images
Page 29 Jay M. Pasachoff/Science Faction/Getty Images
Pages 30-31 SSPL/Getty Images
Pages 32-33 Courtesy of the David Rumsey Map Collection, www.davidrumsey.com
Page 34 Courtesy of the David Rumsey Map Collection, www.davidrumsey.com
Page 35 Courtesy of the David Rumsey Map Collection, www.davidrumsey.com
Page 37 Linda Hall Library of Science, Engineering, and Technology
Pages 38-39 Linda Hall Library of Science, Engineering, and Technology
Pages 40-41 Linda Hall Library of Science, Engineering, and Technology
Pages 42-43 Linda Hall Library of Science, Engineering, and Technology
Page 44 Linda Hall Library of Science, Engineering, and Technology
Page 45 Linda Hall Library of Science, Engineering, and Technology
Pages 46-47 Linda Hall Library of Science, Engineering, and Technology
Pages 48-49 Linda Hall Library of Science, Engineering, and Technology
Page 50 Linda Hall Library of Science, Engineering, and Technology
Page 51 Linda Hall Library of Science, Engineering, and Technology
Page 52 Linda Hall Library of Science, Engineering, and Technology
Page 53 Linda Hall Library of Science, Engineering, and Technology
Page 55 Historical Picture Archive/CORBIS/Corbis/Getty Images
Page 57 DEA/G. CIGOLINI/VENERANDA BIBLIOTECA AMBROSIANA/De Agostini/Getty Images
Pages 58-59 VCG Wilson/Corbis/Getty Images
Pages 60-61 VCG Wilson/Corbis/Getty Images
Page 62 DEA/G. CIGOLINI/VENERANDA BIBLIOTECA AMBROSIANA/De Agostini/Getty Images
Page 63 VCG Wilson/Corbis/Getty Images
Pages 64-65 VCG Wilson/Corbis/Getty Images
Pages 66-67 VCG Wilson/Corbis/Getty Images
Page 68 VCG Wilson/Corbis/Getty Images
Page 69 De Agostini/A. Dagli Orti/Getty Images
Page 70 VCG Wilson/Corbis/Getty Images
Page 71 VCG Wilson/Corbis/Getty Images
Pages 72-73 DeAgostini/Getty Images
Page 74 VCG Wilson/Corbis/Getty Images
Page 75 VCG Wilson/Corbis/Getty Images
Pages 76-77 VCG Wilson/Corbis/Getty Images
Page 78 VCG Wilson/Corbis/Getty Images
Page 79 VCG Wilson/Corbis/Getty Images
Pages 80-81 VCG Wilson/Corbis/Getty Images
Pages 82-83 VCG Wilson/Corbis/Getty Images
Pages 84-85 VCG Wilson/Corbis/Getty Images
Pages 86-87 VCG Wilson/Corbis/Getty Images
Pages 88-89 VCG Wilson/Corbis/Getty Images
Pages 90-91 VCG Wilson/Corbis/Getty Images
Pages 92-93 VCG Wilson/Corbis/Getty Images
Pages 94-95 VCG Wilson/Corbis/Getty Images
Page 96 VCG Wilson/Corbis/Getty Images
Page 97 VCG Wilson/Corbis/Getty Images
Pages 98-99 DEA/G. CIGOLINI/VENERANDA BIBLIOTECA AMBROSIANA/De Agostini/Getty Images
Page 101 Universal History Archive/UIG/Getty Images
Pages 102-103 Fine Art Images/Heritage Images/Getty Images
Page 104 Fine Art Images/Heritage Images/Getty Images
Page 105 DeAgostini/Getty Images
Pages 106-107 DeAgostini/Getty Images
Pages 108-109 DeAgostini/Getty Images
Page 111 SSPL/Getty Images
Pages 112-113 Stapleton Collection/Bridgeman Images
Page 114 Stapleton Collection/Bridgeman Images
Page 115 Historica Graphica Collection/Heritage Images/Getty Images
Pages 116-117 Historica Graphica Collection/Heritage Images/Getty Images
Pages 118-119 Historica Graphica Collection/Heritage Images/Getty Images
Page 120 Stapleton Collection/Bridgeman Images
Page 121 Historica Graphica Collection/Heritage Images/Getty Images
Pages 122-123 Historica Graphica Collection/Heritage Images/Getty Images
Page 125 Courtesy of the David Rumsey Map Collection, www.davidrumsey.com
Page 127 Courtesy of the David Rumsey Map Collection, www.davidrumsey.com
Pages 128-129 JHU Sheridan Libraries/Gado/Getty Images
Pages 130-131 Courtesy of the David Rumsey Map Collection, www.davidrumsey.com
Page 132 Courtesy of the David Rumsey Map Collection, www.davidrumsey.com
Page 133 Courtesy of the David Rumsey Map Collection, www.davidrumsey.com
Page 134 Courtesy of the David Rumsey Map Collection, www.davidrumsey.com
Page 135 Courtesy of the David Rumsey Map Collection, www.davidrumsey.com
Pages 136-137 JHU Sheridan Libraries/Gado/Getty Images
Pages 138-139 JHU Sheridan Libraries/Gado/Getty Images
Pages 140-141 Courtesy of the David Rumsey Map Collection, www.davidrumsey.com
Pages 142-143 Courtesy of the David Rumsey Map Collection, www.davidrumsey.com
Pages 144-145 Courtesy of the David Rumsey Map Collection, www.davidrumsey.com
Page 147 Courtesy of the David Rumsey Map Collection, www.davidrumsey.com
Pages 148-149 Courtesy of the David Rumsey Map Collection, www.davidrumsey.com
Pages 150-151 Courtesy of the David Rumsey Map Collection, www.davidrumsey.com
Pages 152-153 Courtesy of the David Rumsey Map Collection, www.davidrumsey.com
Pages 154-155 Courtesy of the David Rumsey Map Collection, www.davidrumsey.com
Pages 156-157 Courtesy of the David Rumsey Map Collection, www.davidrumsey.com
Page 159 Courtesy of the David Rumsey Map Collection, www.davidrumsey.com
Page 161 Courtesy of the David Rumsey Map Collection, www.davidrumsey.com
Pages 162-163 Courtesy of the David Rumsey Map Collection, www.davidrumsey.com
Page 164 Courtesy of the David Rumsey Map Collection, www.davidrumsey.com
Page 165 Courtesy of the David Rumsey Map Collection, www.davidrumsey.com
Pages 166-167 Courtesy of the David Rumsey Map Collection, www.davidrumsey.com
Page 168 Courtesy of the David Rumsey Map Collection, www.davidrumsey.com
Page 169 Courtesy of the David Rumsey Map Collection, www.davidrumsey.com
Pages 170-171 Courtesy of the David Rumsey Map Collection, www.davidrumsey.com
Pages 172-173 Courtesy of the David Rumsey Map Collection, www.davidrumsey.com
Pages 174-175 Courtesy of the David Rumsey Map Collection, www.davidrumsey.com
Page 176 Courtesy of the David Rumsey Map Collection, www.davidrumsey.com
Page 177 Courtesy of the David Rumsey Map Collection, www.davidrumsey.com
Page 179 Magruder, Kerry/Share Ok
Page 180 Dominio pubblico, Courtesy of the Library of Congress's Prints and Photographs division
Page 181 Dominio pubblico, Courtesy of the Library of Congress's Prints and Photographs division
Page 182 Dominio pubblico, Courtesy of the Library of Congress's Prints and Photographs division
Page 183 Dominio pubblico, Courtesy of the Library of Congress's Prints and Photographs division
Page 184 Dominio pubblico, Courtesy of the Library of Congress's Prints and Photographs division
Page 185 Dominio pubblico, Courtesy of the Library of Congress's Prints and Photographs division
Pages 186-187 Dominio pubblico, Courtesy of the Library of Congress's Prints and Photographs division
Pages 188-189 Dominio pubblico, Courtesy of the Library of Congress's Prints and Photographs division
Pages 190-191 Dominio pubblico, Courtesy of the Library of Congress's Prints and Photographs division
Pages 192-193 Dominio pubblico, Courtesy of the Library of Congress's Prints and Photographs division
Pages 194-195 Dominio pubblico, Courtesy of the Library of Congress's Prints and Photographs division
Page 197 Courtesy of the David Rumsey Map Collection, www.davidrumsey.com
Page 199 Courtesy of the David Rumsey Map Collection, www.davidrumsey.com
Page 200-201 Courtesy of the David Rumsey Map Collection, www.davidrumsey.com
Page 202-203 Courtesy of the David Rumsey Map Collection, www.davidrumsey.com
Page 204-205 Courtesy of the David Rumsey Map Collection, www.davidrumsey.com

Text

Elena Percivaldi

Project editors

Valeria Manferto De Fabianis

Laura Accomazzo

Graphic design

Paola Piacco

WS White Star Publishers® is a registered trademark
property of White Star s.r.l.

Piazzale Luigi Cadorna, 6 - 20123 Milan, Italy
www.whitestar.it

Translation: Richard Pierce

ISBN 978-88-544-1310-8
1 2 3 4 5 6   22 21 20 19 18

Printed in Italy by Rotolito - Seggiano di Pioltello (Milan)

Ursa Major
Coma Berenices
Leo
AUTUMNAL COLURE
Denebola
Virgo
ECLIPTIC
SEPTEMBER